INEXPENSVE
3D PRINTER
PROJECTS

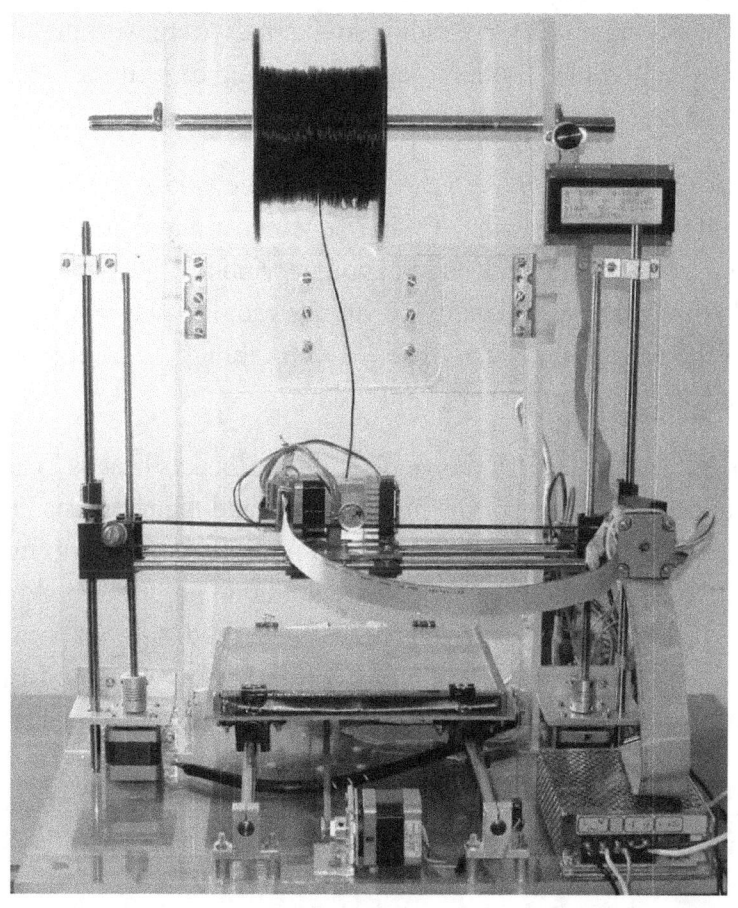

Robert J Davis II

Update: This is an updated version that adds compatibility with the Prusa i3 and the Geeetech i3 3D printers. Parts were interchanged between these models to verify that they are compatible.

Dedication: I thought I would dedicate this book to my stepfather Charles L Nichols. He encouraged me when I was a teenager to "play" with electronics. At one point we had a shop set up in the basement of his house in Athens Pennsylvania with several TV's and at least three antique radios. I had fixed up all three radios so that they were working. Back then the biggest problem with radios was dried up capacitors. Then the flood of 1972 hit and filled the basement with mud.

Charles taught me how to use a cardboard device where you could line up a picture of the problem with a TV in one window and then you could see what tubes needed to be replaced in another window. He once helped me to get a "pulpit" that was actually an antique radio to play with. He helped me to get the various parts that I needed to fix things. His help was instrumental in me being where I am today.

Disclaimer: The projects in this book are described as best as I can. They can be dangerous because of the high currents and temperatures. However the builder assumes all responsibility and any liability for anything that he builds. If you do not know what you are doing do not try to build this yourself. If this book is "over your head" you could start out with simpler projects like those that are found in my earlier books "Arduino LED Projects", "Arduino LCD Projects" or "Arduino Robotics Projects".

This book is intended to help you to be able to build your own 3D printer. Feel free to improve upon my design or even come up with your own designs and improvements. I will show you how to make almost everything yourself. To be honest purchasing and assembling a kit is a much easier solution.

Table of Contents

RepRap Pololu Shield Wiring.
X, Y and Z Limit switches.
Heat Bed wiring and power supply.

Teacup Firmware
Arduino IDE
Pronterface PC Software
Slic3r (3D Slicer)
Upgrading to Marlin Firmware

X, Y, Z Calibration
Extruder Calibration
Making changes to config.h
Making things with PLA and ABS

Left side spool holder
Extended gantry braces spool holder

Making custom cables
Rotary encoder
Modifying Configuration.h

Introduction

To 3D Printing

Making a 3D printer can be a very expensive and complicated project. I have been building electronics for over 40 years and this is by far the most complex thing that I have ever attempted to build. Making this 3D printer should not be your first project. You will either need to have some experience in building things or a mentor that has lots of experience and who has the time to help you out.

This project is designed with some of the same emphasis that my "Inexpensive CNC projects" book and design had.

1. KISS as in "Keep It Simple Stupid", try to make the complex things look easy.

2. Keep it inexpensive, almost no one has thousands of dollars to spend on their projects.

3. This design will not require a CNC mill or a 3D printer to make it as much as is possible. (I did buy some printed parts)

4. Use standard "off-the-shelf" parts as much as is possible. Use parts from local hardware stores when it is possible to do.

That being said I did finally have to give in and buy the parts for the X-axis ends to get started. They are just too complicated to be easily made without a 3D printer. I do show how you can use X-axis ends for a Geeetech i3 that are commonly available on eBay for $25.

What is a 3D printer? Imagine if you took a hot melt glue gun, and added a really fine tip to it. That way you can make really small drops of glue.

Then install it into a CNC machine so that you can place the glue drops in any very precise X, Y, and Z coordinate. Then add a feed rate control system so that you can control how fast the raw hot melt glue goes into the glue gun. You now have a simplified explanation of how this machine works. In reality the "glue sticks" are very small in diameter and continuous.

The raw material used in a 3D printer is called a "Filament", it is much smaller in diameter than what a glue gun uses. There are two common sizes, 3 mm and 1.75 mm. The smaller diameter filament can be heated faster and used with smaller extruder parts. A smaller nozzle size means that you have more precise plastic control so that the end product looks more professional.

Printer filament is available in a number of materials. The two most common materials are called PLA (Polylactic Acid) and ABS (Acrylonitrile Butadiene Styrene). Here is a chart comparing the two materials.

PLA

Extrudes at around 180 degrees C
Heat bed is an option
Helps it to cool off while printing
You can use bare glass or painters tape
Tends to curl on the corners
Doesn't produce a bad smell
Plant based, biodegradable

ABS

Extrudes at around 220 degrees C
Requires a Heat bed
Does not need to cool while printing
Works best with Krapton tape
Tends to warp and crack

Must be ventilated because of the smell
Oil Based

Another problem in making this 3d printing machine was to decide if I should model it after the Prusa i3, the Geeetech i3, or after the Mendel 90. The original Prusa 3D printer had a lot of issues and it looked like an ugly pile of bolts. Several improved versions of the Prusa have been made. The Prusa Air is dimensionally like the Prusa but the side bolts and top bolts are replaced with a Plexiglas frame. The Mendel 90 took it a step further and replaced all the bolts with Plexiglas or wood for the sides and bottom.

The Mendel 90 has two variations too. The Mendel 90 "Sturdy" uses 10 mm rods and bearings. The Mendel 90 "Dibond" used 8 mm rods and bearings. The Mendel 90 looks better but the 8 mm rod and bearing version uses 5 mm lead screws instead of 8 mm! Why they made the change I do not know. It might have been to get more threads per inch without driving the price up. Also the Z axis guide rods collide with standard motor mounts. The only solution to that is to use some custom made motor mounts!

The Prusa Air and the Mendel 90 have faded as the Prusa i3 and Geeetech i3 have become the most common 3d printers. So my Mendel 90 compatible printer has now been upgraded to an i3.

At first I built a combination of the Mendel 90 and the Air. It had the framework like that of the Mendel 90 but the leadscrew will be 8mm and the leadscrew to guide rod spacing will be that of the Prusa Air making motor mounts easier to make.

To make sure this 3D printer is similar in design to the Mendel 90, I printed out the pdf files of the framework and taped them together. The width was supposed to be 18.4 inches but my paper pieces were 18.2 inches wide when I taped them together. My base that I am using is only 18 inches wide so the right .2 or .4 inches will have to be skipped. This whole idea of doing the framework by printing papers that are taped together is kind of crazy! I greatly prefer to use normal mechanical drawings, and that those drawings are in English!

The work area of the 3D printer will be an "X" of 8 inches by a "Y" of 8 inches by a "Z" of 8 inches in size. This is the standard 3D printer size. It requires a gantry that is 15 inches high.

Before you attempt to build your own 3D printer I highly recommend that you look at every 3D printer design that you can find on the Internet to get some ideas. I even downloaded dozens of pictures and looked through them many times. I tried to pick out what features that I wanted to include in this 3D printer.

Just to make sure that we are all on the same page, the "X" axis moves from the right to the left, the "Y" axis moves from front to back and the "Z" axis moves up and down. There is an additional extruder stepper motor that is used for controlling the feed rate of the filament.

Besides controlling the X, Y, Z and the extruder stepper motor the machine will also need to have some temperature controllers. They are needed for the extruder hot end temperature and for the hot bed temperature. All of these things can be controlled with an Arduino Mega processor and a RAMPS board. The RAMPS adapter board is needed to interface the Mega to everything on the 3d printer. That will be covered in the electronics section of this book.

Chapter 1

Design Concepts

Here is a drawing showing the front view of what I envisioned it would look like. The X axis ends have to be swapped around otherwise the stepper motor would have to go through the left side of the gantry. There is another design that has the motor inside of the X axis end, instead of sticking out of it. The Prusa i3 has a rotated X axis, so the stepper motor is located on the left side but the gantry is narrower.

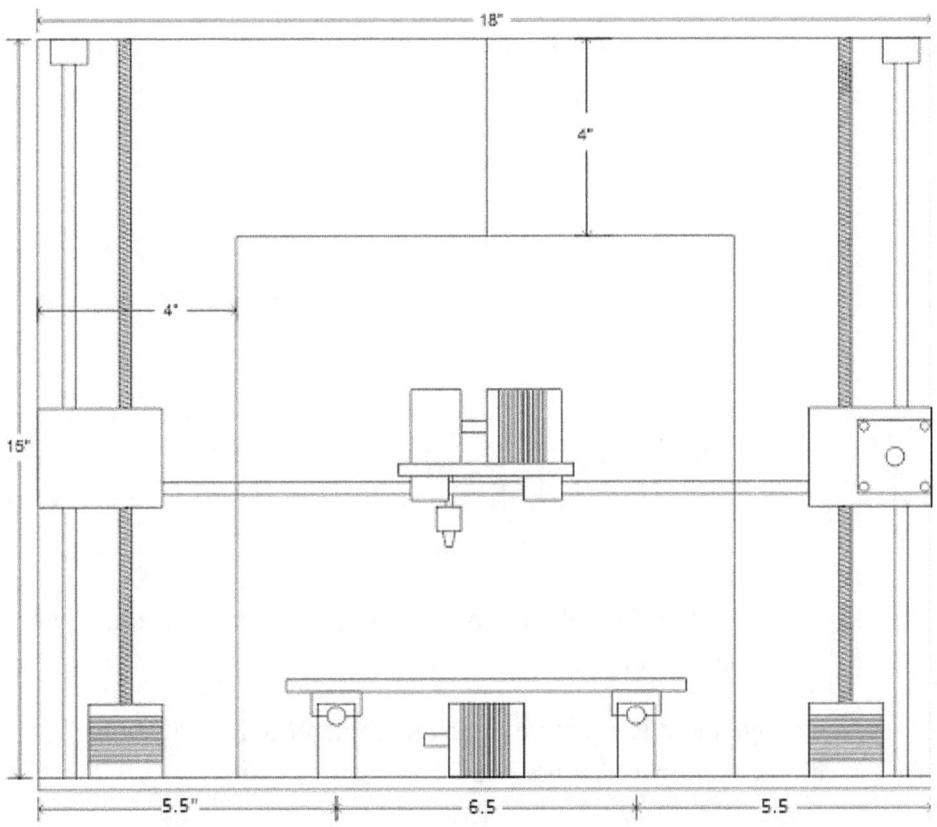

The correct spacing of the Y axis guide rods for the Prusa i3 is 6.75 inches to be exact. The sides of the Prusa i3 Gantry are only 2.125 inches wide instead of four inches wide like in the Mendel 90.

Here is a drawing that shows the side view of the 3D printer. The gantry braces are angled because I made the gantry out of two pieces of 24 inch by 8 inch Plexiglas. As a result the gantry sides needed to be fastened together in the middle and the back pieces needed to be cut on an angle to hide where the gantry pieces were cut out of them. The gantry needs to be 7 inches from the back of the machine.

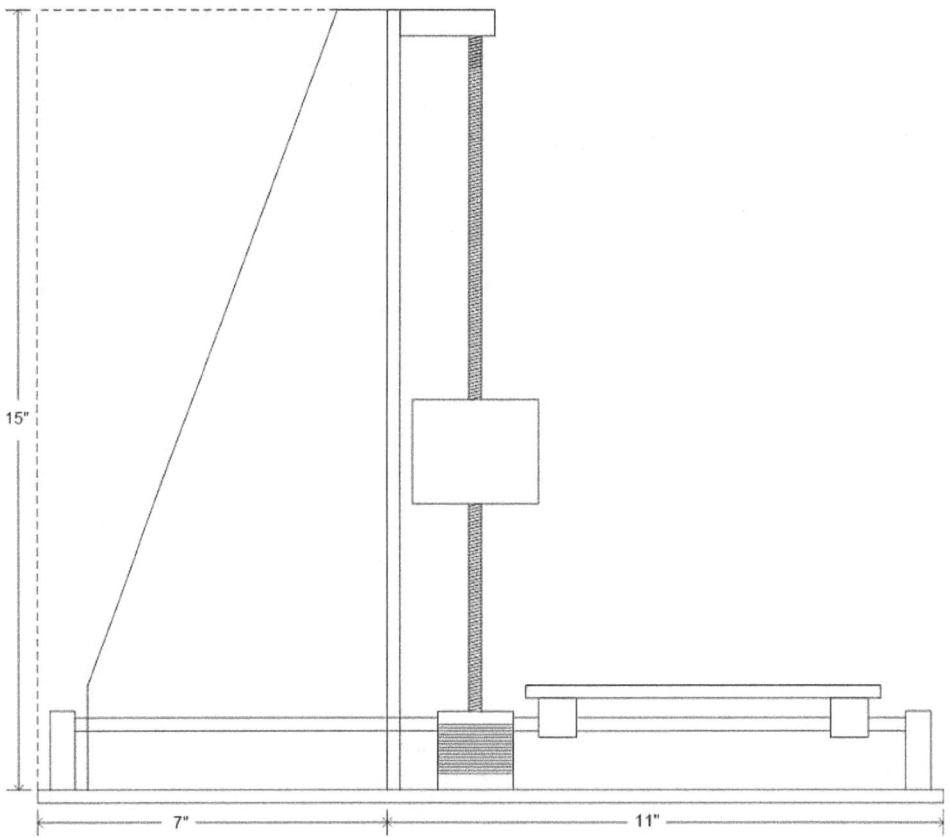

These drawings do not show the motor mounting brackets and some other smaller mechanical details.

On the next page there is a picture of an assembled Mendel 90 style of 3D printer with the main parts pointed out.

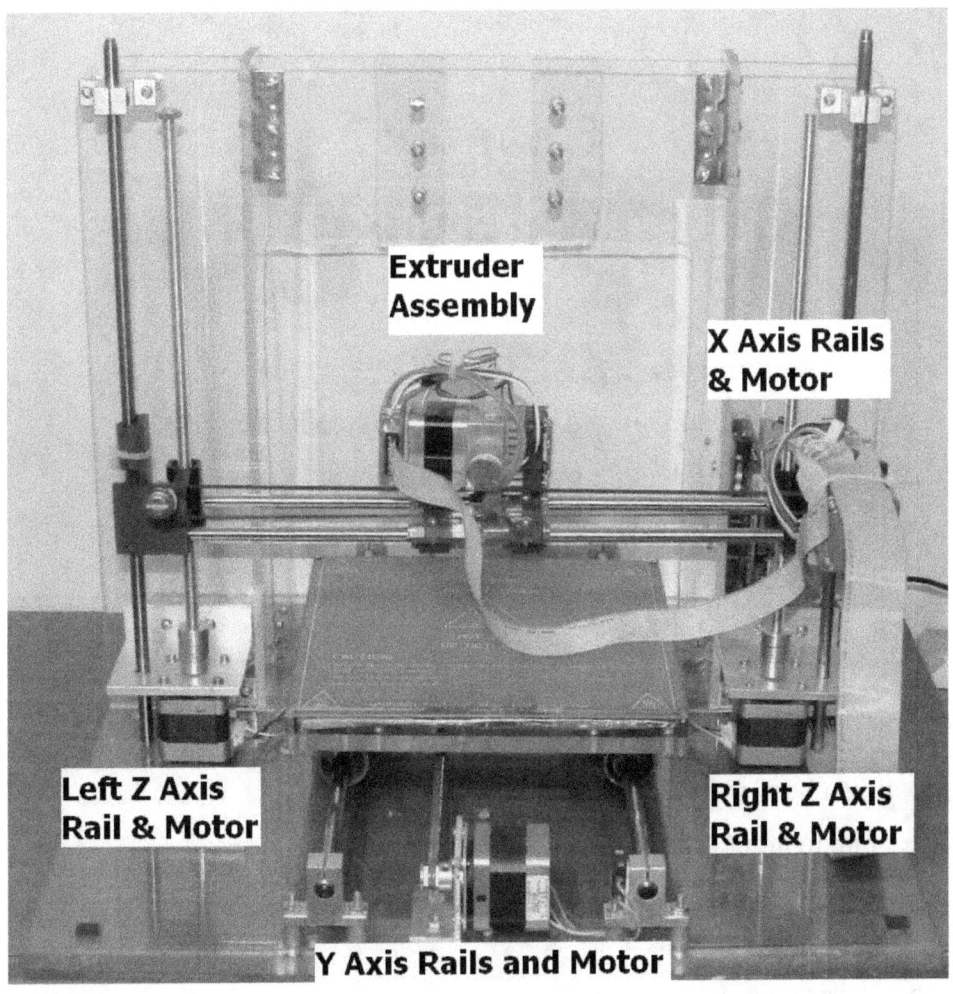

Extruder Assembly

X Axis Rails & Motor

Left Z Axis Rail & Motor

Right Z Axis Rail & Motor

Y Axis Rails and Motor

The Prusa and Geeetech i3 has the X axis stepper motor located on the left side and behind its mounting bracket. They also have the X axis rails located one above the other instead of side by side.

Chapter 2
Bill of Materials (BOM)

Rods, holders and bearings:

2 x 18 inches (450mm) 10mm diameter smooth rods.	$15
2 x 15.75 inches (400mm) 8 mm diameter smooth rods.	$15
2 x 12 inches 5/16 inches or 8mm diameter threaded rods.	$ 2
4 x 10 mm rod holders.	$13
4 x 10 mm linear bearings.	$ 8
4 x 10 mm bearing holders (can be printed).	$ 6
8 x 8 mm linear bearings.	$ 6
4 x 8 mm bearing holders.	$ 2
2 x 5mm to 8mm flexible couplings	$ 6
2 x 16 or 20 teeth GT2 Timing Pulley	$ 6

Plexiglas or Plywood:

18" by 18" by 3/8 inch thick base.

2 x 15" by 9" gantry sides. (or one 18" by 15")

2 x 15" by 6" gantry back braces.

Hardware:

4 x 2.5 inch metal corner braces.	$ 4
4 x 1/4 inch bearings (other bearing options in text)	$ 8

40 x 8-24 by 3/4 inch long screws and nuts.

8 x 4-40 screws and nuts for mounting the Arduino and the Head bed.

20 x 3mm screws for mounting the stepper motors.

4 x one inch spacers for the Z axis rods.

4 x 3/4 inch spacers for the Y axis rods.

4 x 3/8 inch spacers (or springs) for mounting the heat bed.

Electronics:

5 x NEMA 17 stepper motors	$50
1 x Reprap shield with driver boards	$20
1 x Arduino Mega	$14
2 x 100K Thermistor	$ 1
1 x Heat bed	$10
3 x Mechanical end stop switches	$10
1 Extruder Assembly / Print Head	$50

12 Volt 5 Amp AC Adapter. (From junk pile)

12 Volt 12 Amp Power Supply (From junk pile)

Many 2 and 4 pin connectors (From old TV/Monitors)

15 or 9 pin Male and Female connectors (optional) $ 5

Printed parts:

X Axis ends, or Prusa Printed parts, or Geeetech kit $30

Home made parts:

2 x NEMA 17 motor mounts with hole for guide rods.

1 x NEMA 17 low profile motor mount

TOTAL APPROX: $300

Chapter 3
Parts Sources

The 8 mm smooth rods were found on eBay. Most of the rods I used came from printers and scanners that I have stripped for parts. These rods are really hard and difficult to cut shorter. It is best to get them from one vendor that way they are the same length.

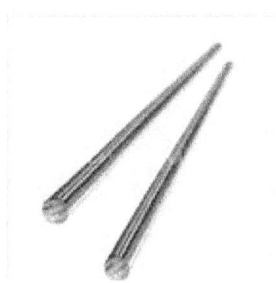

1pc OD 8mm x 400mm $5.49
Bearing Steel Cylinder Liner   + $1.99
Rail Line... (371097841987)   shipping
Sale date: 09/09/14
Tracking number: --
Estimated Delivery: Fri. Sep. 19 -
Mon. Oct. 6

The 8 mm rod holders were found on eBay (Ignore this price). I won a 1 cent auction for the 8mm rod holders.

2pcs SK8 8mm 0.3" CNC $0.01
Linear Rail Shaft Guide   Free shipping
Support (390919533062)
Sale date: 08/28/14
Tracking number: LK171481705CN
Estimated Delivery: Thu. Sep. 11 -
Thu. Sep. 25

Lm8UU Linear bearings were also found on eBay (You only need 8 if you have 10mm rails for the Y axis). Skimping here can be a problem. My bearings do not move very smoothly at all.

10pcs LM8UU 8mm Linear Ball Bearing
Bush Bushing
(321354645930)
Sale date: 08/20/14
Tracking number: --
Estimated Delivery: Wed. Sep. 3 - Wed. Sep. 17

$6.99

Free shipping

The SK 10 mm rod holders as they were found on eBay. You will need four of these for the optional 10mm Y axis.

2x SK10 10mm CNC Linear Rail Shaft Guide
Support MPH (321320931642)
Quantity: 2
Sale date: 08/20/14
Tracking number: LK162618431CN
Delivered: Thu. Aug. 28

$6.42

Free shipping

The 10mm linear bearing as they are found on eBay. You will only need four for a 100 Y axis. These bearings work great!

10pcs LM10UU 10mm
10x19x29mm Linear Ball
Bearing Bush Bus...
(121170026097)
Sale date: 03/17/14
Tracking number: LK986692994CN
Delivered: Mon. Mar. 31

$8.55

Free
shipping

The GT2 Timing gears are found on eBay. These should have been 20 teeth pulleys. I had to make changes in config.h to compensate for having the wrong gears. The steps per mm is set to 100 for 16 teeth gears.

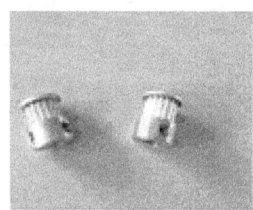

2 Pcs GT2 16 Tooth 5mm Bore Aluminum Timing
Pulley For 3D... (301246377963)
Sale date: 08/22/14
Tracking number: LK167097696CN
Delivered: Fri. Aug. 29

$4.49

+ $1.50
shipping

The GT2 - 20 teeth gears use a setting of 80 steps per mm.

3D Printer GT2 20 Teeth Aluminum Timing Belts 5mm Shaft
Pulleys CNC, Silver T1
(252603844952)

Choose: 2 pcs

ITEM PRICE:
US $1.38

ⓘ Estimated delivery **Tue, Feb 21 - Thu, Mar 09**

The Z axis Flexible couplings found on eBay. These were drilled a little crooked!

2X Flexible Z Axis Coupler 5mm*8mm 5/16"
RepRap 3D Printe... (171393972451) $3.96
Sale date: 08/22/14   + $1.50
Tracking number: LK167087032CN   shipping
Delivered: Fri. Aug. 29

The 100K Thermistors are from eBay. USPS sent these all over the USA and I never received them. The second order arrived with no problem.

2pcs 100K ohm NTC Thermistor for RepRap
heatbed MK2a Prus... (281418790783) $0.99
Sale date: 08/21/14   Free shipping
Tracking number: LK166742403CN
Estimated Delivery: Wed. Sep. 3 - Thu. Sep. 18

The X Axis end pieces were found on eBay. They do not actually match what is normally used in a Mendel 90 design but they did work.

Prusa i2 LM8UU X-Ends (151075320571) $12.00
Type: Metric
Sale date: 08/20/14   + $3.22
Tracking number: 9400109699939172486398 shipping
Delivered: Sat. Aug. 23

The Prusa i3 printed parts can be found on eBay or you can print your own once you have a working printer.

RepRap Prusa i3 Rework Printed Parts Kit - HIGH QUALITY GLOSSY BLACK PLA

USA MADE & IN STOCK for FAST Shipment A++ Quality

★ ★ ★ ★ ★ 1 product rating

$29.99

Buy It Now

Free shipping

302 sold

If you want to buy an i3 frame you can do that rather than make your own frame. There are really cheap ones available from overseas.

3D Printer Reprap Mendel Prusa i3 Frame Laser Cut 6mm MDF + Screws (281892769162)

ITEM PRICE: US $17.99

The Arduino Mega was found on eBay. I had one but it would no longer communicate so I bought a new one and it worked perfectly.

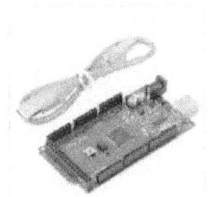

ATmega2560-16AU CH340G MEGA 2560 R3 Board + USB Cable for Arduino B2 USA (122288607349)

$ 🖥 ◀ ▮ ▰

Delivered on **Wed, Feb 08**

Tracking number: 9374869903501742906262

ITEM PRICE: US $7.35

The Ramps Control board with the five included stepstick driver boards was purchased on eBay.

RAMPS 1.4 Control Board + 5X A4988 Stepstick Driver Modul... (201137444335)

Sale date: 08/15/14

Tracking number: RI047961227CN

Estimated Delivery: Wed. Aug. 27 - Fri. Sep. 12

$19.00

Free shipping

The NEMA 17 Stepper motors as on eBay. Here are two examples. The first ones have sound dampening devices built in. The minimum power rating is around 40 oz per inch.

3 X Nema 17 Japan Servo Stepper Motor 44oz/in Robot RepRap Makerbot Prusa 3D
(181118082938)

ITEM PRICE
US $25.00

Here is a second example of some NEMA 17 stepper motors available on eBay.

Lot of 5 Stepper motors NEMA 17- 44oz/in CNC ROUTER ROBOT REPRAP MAKERBOT Prusa

$39.97
Buy It Now
+$10.00 shipping

The Head Bed as found on eBay. It arrived broken in one corner but it did work fine anyway.

3D Printer PCB Heatbed MK2a For Prusa Mendel 3D printer variants latest TR
(181293243647)

ITEM PRICE
US $9.03

There is an optional metal Heat bed available. It was bent in the mail but clamping it to the glass will straighten it out. It does not come with wires.

RepRap 3D Printer Heatbed MK2B 12V/24V PCB Hot Plate Heat Bed For Prusa Mende BE
(201778959832)

ITEM PRICE:
US $6.58

Delivered on **Mon, Feb 27**
Tracking number **LK161817970CN**

The Print Head or extruder was purchased on eBay. This is actually the wrong part. Usually a Mendel 90 uses a "Wades" geared extruder with 3 mm filament. This extruder requires 1.75mm filament. Its feed mechanism was not that great either, it slipped way too much.

0.4mm Nozzle Extruder Print Head for 3D Printer   ITEM PRICE
Makerbot Mendel RepRap   US $55.55
(201104592001)

You might need 9 and or 15 and or 25 pin male and female connectors. I had some in stock but I bought a few more.

5PCS Parallel Port Connector DB25 25-Pin Adapter   ITEM PRIC
Female   (251074888411)   US $2.94

I also bought a big spool of PLA filament that has lasted a long time.

12 Colors New 3D-Printer Filament 1.18kg/2.6lb PLA 3mm or   Free shipping
1.75mm Repraper Reprap
(141025862698)

Size: 1.75mm   ITEM PRICE:
Color: Black   US $17.99

The Corner braces can be found at Lowes.

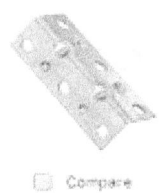

Stanley-National Hardware 2-1/2-in Zinc Corner $2.98
Brace
★ ★ ★ ★ ★ (1 Review)   Qty.: 1

Item #: 315639 | Model #: DFB113   Add to Cart +

- Manufactured from steel
- Product design allows for quick and easy repair of general household items
- Used for reinforcing inside of right angle corner joints

The 1/2 inch pipe holders that can optionally be used for the 8mm bearings are from Lowes.

AMERICAN VALVE 5-Pack 1/2-in Plastic Standard Clamps $2.32

★ ★ ★ ★ (2 Reviews)   Qty.: 1

Item #: 301296   Model #: AV301296   Add to Cart +

- Used to secure CTS piping in residential and commercial applications
- Ribs allow pipe to glide easily and silently when expanding and contracting
- Split design allows for use after piping has been installed

☐ Compare

The 1/4 inch bearings with a flange can be bought from eBay and they are used for the belt idlers. These are much better than the standard bearings.

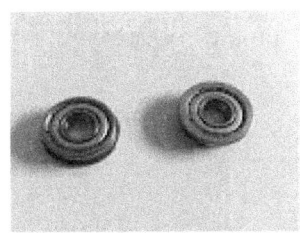

2pcs Miniature Flanged Balls Bearing High Precision FR4ZZ 1/4"x5/8"x0.196" inch

$4.00   From China

Buy It Now

Free shipping

The 5/16 by 12 inch threaded rods that I used were from Lowes. They only fit the Mendel 90 clone.

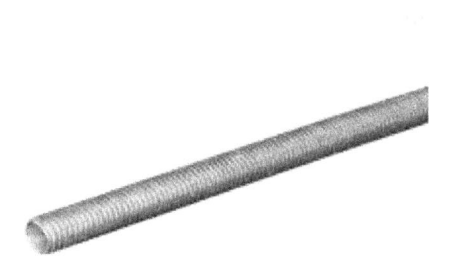

Steelworks 5/16-in x 12-in Standard (SAE) Threaded Rod

Item #: 44591 | Model #: 11012.0

★ ★ ★ ★ ★ ☑ 1 review | Write a review

$1.19

The timing belt can be found on eBay. I ordered 6 feet; that would be more than enough. Then I also used the belts that I salvaged from some old scanners.

GT2 Timing Belt by the foot 6mm width,3D printer Rostock Mendel REPRAP $ 🛒

(330967958173)   ITEM PRICE

Quantity: 6   US $5.94

Here is a gear that can be used to make a better extruder. It is curved to better match the shape of the filament.

New MK7 Stainless Steel Extruder Drive Gear Hobbed Gear   ITEM PRICE:
For Reprap 3D Printer   US $1.72
(261567139009)

I also bought some high temperature Kapton tape for use with the PLA filament.

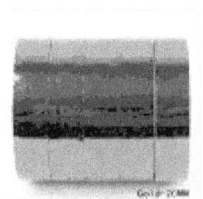

PM 1pcs 100ft High Temperature Heat Resistant Polyimide   ORDER TOTAL
Kapton Tape Gold 20mm   US $1.27
(131326974052)
                                                          Free shipping

                                                          ITEM PRICE:
                                                          US $1.27

To upgrade to the Metal Geeetech I3 you will need the X axis ends and 8mm leadscrews. I purchased two of the leadscrews that came with the stepper motors, nuts and couplers. The steppers even came with nice long wires that had the needed 4 pin plugs at their ends.

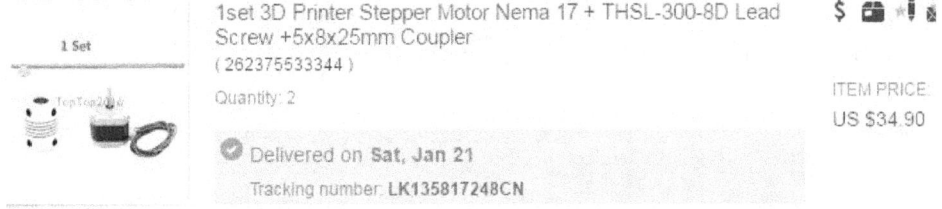

1set 3D Printer Stepper Motor Nema 17 + THSL-300-8D Lead
Screw +5x8x25mm Coupler
(262375533344)

Quantity: 2   ITEM PRICE:
                                                          US $34.90

Delivered on Sat, Jan 21
Tracking number: LK135817248CN

The Geetech metal X axis does not say that it is the X axis with the carriage. It also fails to mention that it needs the 8mm leadscrews to work. With shipping it is $25 for the kit.

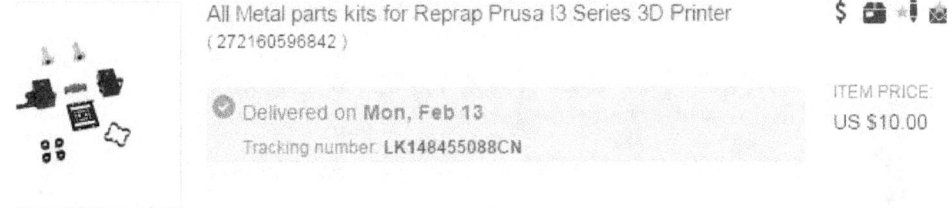

All Metal parts kits for Reprap Prusa I3 Series 3D Printer
(272160596842)

                                                          ITEM PRICE:
Delivered on Mon, Feb 13   US $10.00
Tracking number: LK148455088CN

You only need 3 limit switches but if you do not have some micro switches sitting around you can buy some. These come with LED's to light when activated and nice long wires to the needed 3 pin plugs.

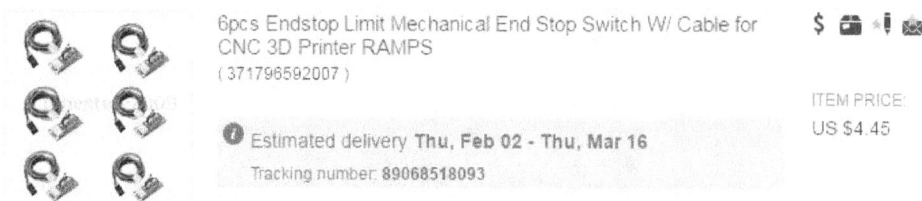

6pcs Endstop Limit Mechanical End Stop Switch W/ Cable for CNC 3D Printer RAMPS
(371796592007)

$ 🔲 ★📱 ✉

ITEM PRICE:
US $4.45

ℹ Estimated delivery **Thu, Feb 02 - Thu, Mar 16**
Tracking number: 89068518093

If you are not upgrading to the Geeetech, but want a metal X axis carriage this one works nicely.

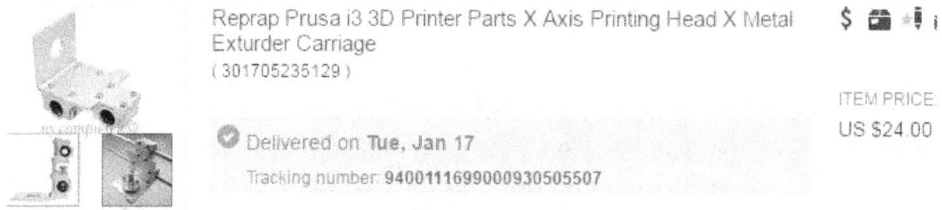

Reprap Prusa i3 3D Printer Parts X Axis Printing Head X Metal Exturder Carriage
(301705235129)

$ 🔲 ★📱 ℹ

ITEM PRICE:
US $24.00

✔ Delivered on **Tue, Jan 17**
Tracking number: 9400111699000930505507

You can also purchase a metal Y axis carriage but it only supports using three bearings and 8mm rods. I prefer using four 10mm bearings.

Prusa i3 Universal Y Carriage Plate Upgrade, Aluminum Anodized for 3D Printer
(121561406955)

$ 🔲 ★📱 ✖

ITEM PRICE:
US $18.89

✔ Delivered on **Sat, Jan 14**
Tracking number: 9405511699000322755544

Another option is to buy a metal motor mount for the Y axis motor. This can also be used to mount the extruder.

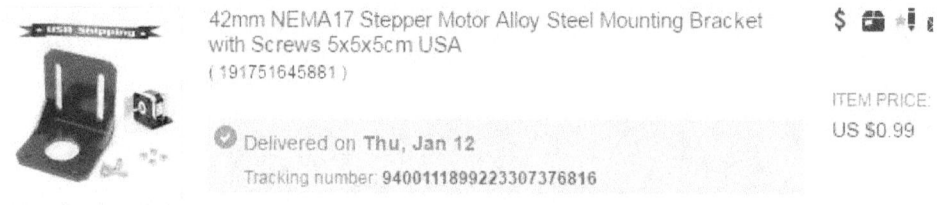

42mm NEMA17 Stepper Motor Alloy Steel Mounting Bracket with Screws 5x5x5cm USA
(191751645881)

$ 🔲 ★📱 ▪

ITEM PRICE:
US $0.99

✔ Delivered on **Thu, Jan 12**
Tracking number: 9400111899223307376816

Chapter 4
Mechanical Drawings

Building this 3D printer will require a lot of metric to inches and inches to metric conversions. It is best to get a ruler that is marked in both units of measurement. I use my computer sometimes to convert back and forth. All of my drawings will be in inches.

Here are the NEMA 17 Stepper motor specs converted into inches. This drawing is from Reprap.org, but I have converted everything to inches. These numbers will be needed to make the motor mounts.

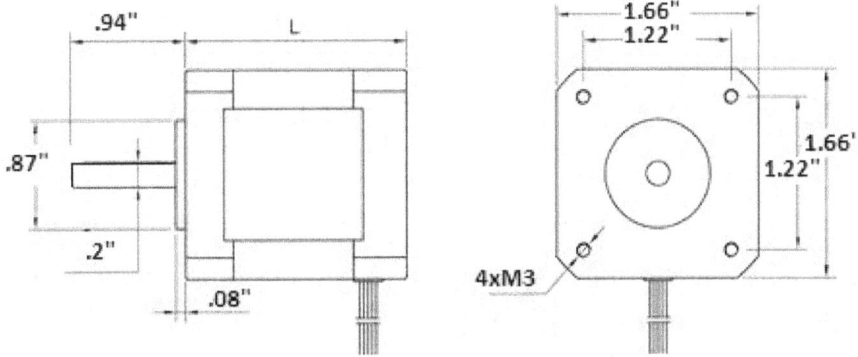

Here are the linear Bearing specifications. All dimensions are still in mm. Both LM8UU and LM10UU bearings were used in making this machine.

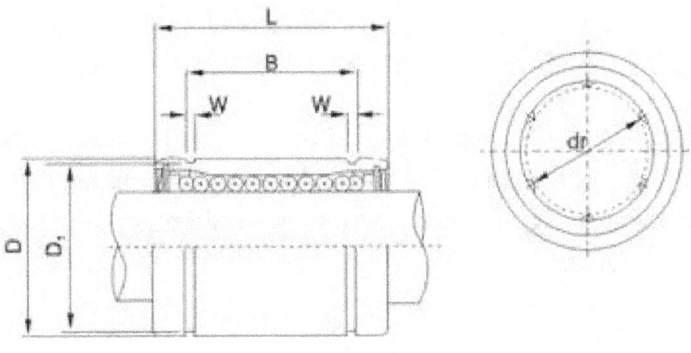

Type	D – Out. Dia.	L- Length	B – Rings	W – Width
LM8UU	15mm	17/24mm	11.5/17.5mm	1.1mm
LM10UU	19mm	29mm	22mm	1.3mm

Here are the rail or rod holder specifications. All dimensions are still in mm. Both SK8 and SK10's were used in making this 3D printer, but you could just stick to using 8mm rails and bearings.

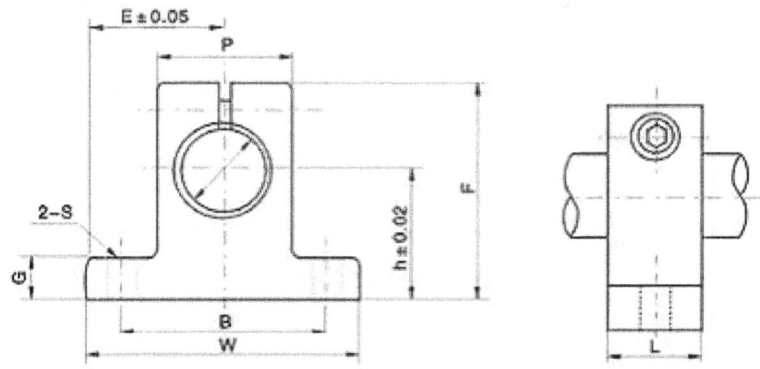

Model	Shaft Dia.	H	E	W	L	F	G	P	B	Mtg. Bolt
SK8/10	8/10	20	21	42	14	37.5	6	18	32	M5

I had an 18 inch by 18 inch by 3/8 inch piece of Plexiglas that I used for making the 3D printer base. It had been used previously to make a vacuum tube Tesla coil at one time so it had a few extra holes in it.

Then I purchased some 24 inch by 8 inch by 1/4 inch pieces of Plexiglas at a yard sale for only $1 each. I needed to make the gantry and the gantry braces out of just two of those pieces. After several drawings and a great deal of figuring here is the cutting pattern that I came up with.

Note that this design has factory edges for the outside of the gantry and for the inside of the gantry braces. I was making my cuts with a jig saw and needed the accurate factory cuts where they would make sure that the machine was square.

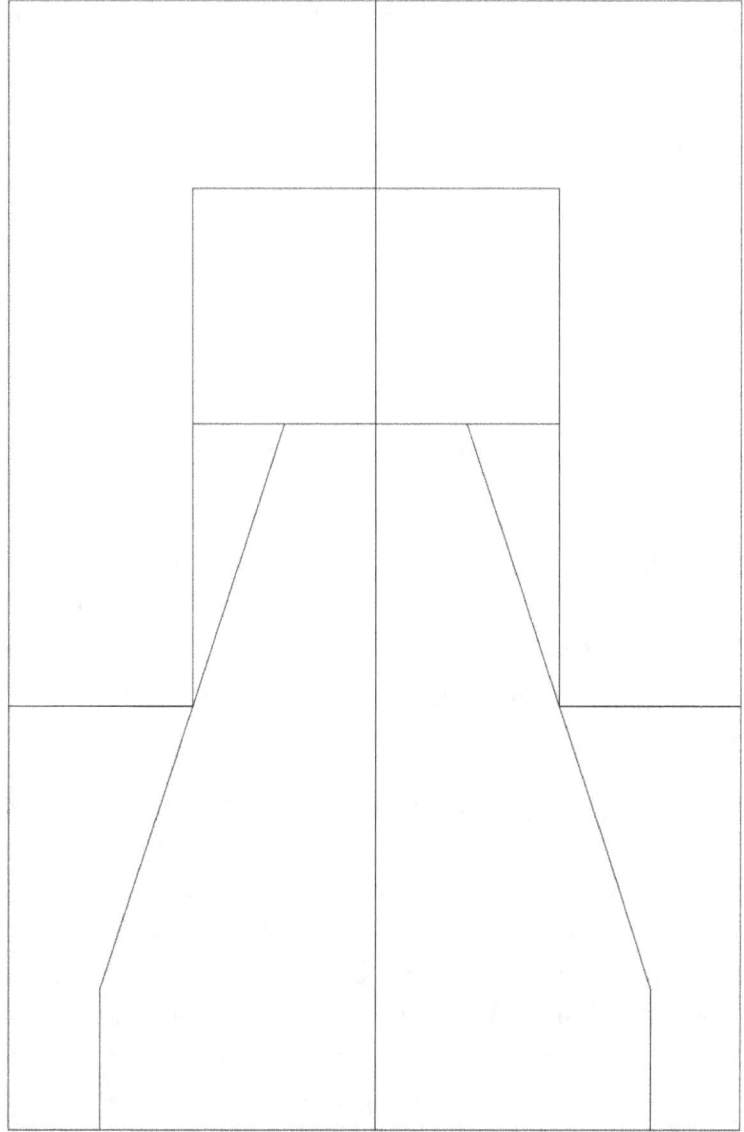

The next few drawings are presented with limited text in order to use the entire page for the drawing thus making them as large as is possible. On the next few pages there is the mechanical drawing of the base, the left side is shown, the right side is symmetrical except for the Y motor mount and belt bearing holder. These drawings are less than 1/2 size in scale in order to fit them into this book. The actual size of the base is 18 inches by 18 inches.

After that there is the Gantry mechanical drawing of the left side, the right side is symmetrical. The gantry is 18 inches wide by 15 inches tall for the Mendel 90 version. Then there is the Prusa i3/Geeetech i3 gantry drawing. Once again it is symmetrical.

Then there is the drawing of the gantry braces. The can be rectangular at about 15 inches tall by six inches wide. Or you can angle them like I did to save on the use of Plexiglas.

All holes are clearance holes for 8-32 screws unless marked otherwise. The size I used was 5/32 inch or about .16 inches in diameter. I sometimes had to "ream" them with a file or the drill just a little in order to fit the 8-32 screws.

An issue with these drawings is that 1/2 inch is ".5", 1/4 inch is ".25", 1/8 inch is ".125" (usually it rounds up to .13) and 1/16 inch is ".0625" (usually it rounds off to .06). However the CAD program that I am using sometimes rounds the wrong way and as a result the numbers do not add up correctly.

My spacing between the Y axis rails was 6.25 inches. I think in the Mendel 90 design they are closer to 6.5 inches apart. The Prusa i3 uses 6.75 inch Y axis rail spacing. The base hole patterns are for the Mendel 90 design with the stepper motor located at the front, the Geeetech and Prusa i3 have the stepper motor at the back. In that case swap their mounting hole positions. Also the i3 designs do not need nearly as many holes to fasten the gantry to the base. I used one 1" by 1" L bracket fastened to the bottom motor mounting screw and then to a single hole in the base.

IMPORTANT NOTE: This hole pattern is for the modified Mendel 90 design. The i3 designs do not need nearly as many holes!

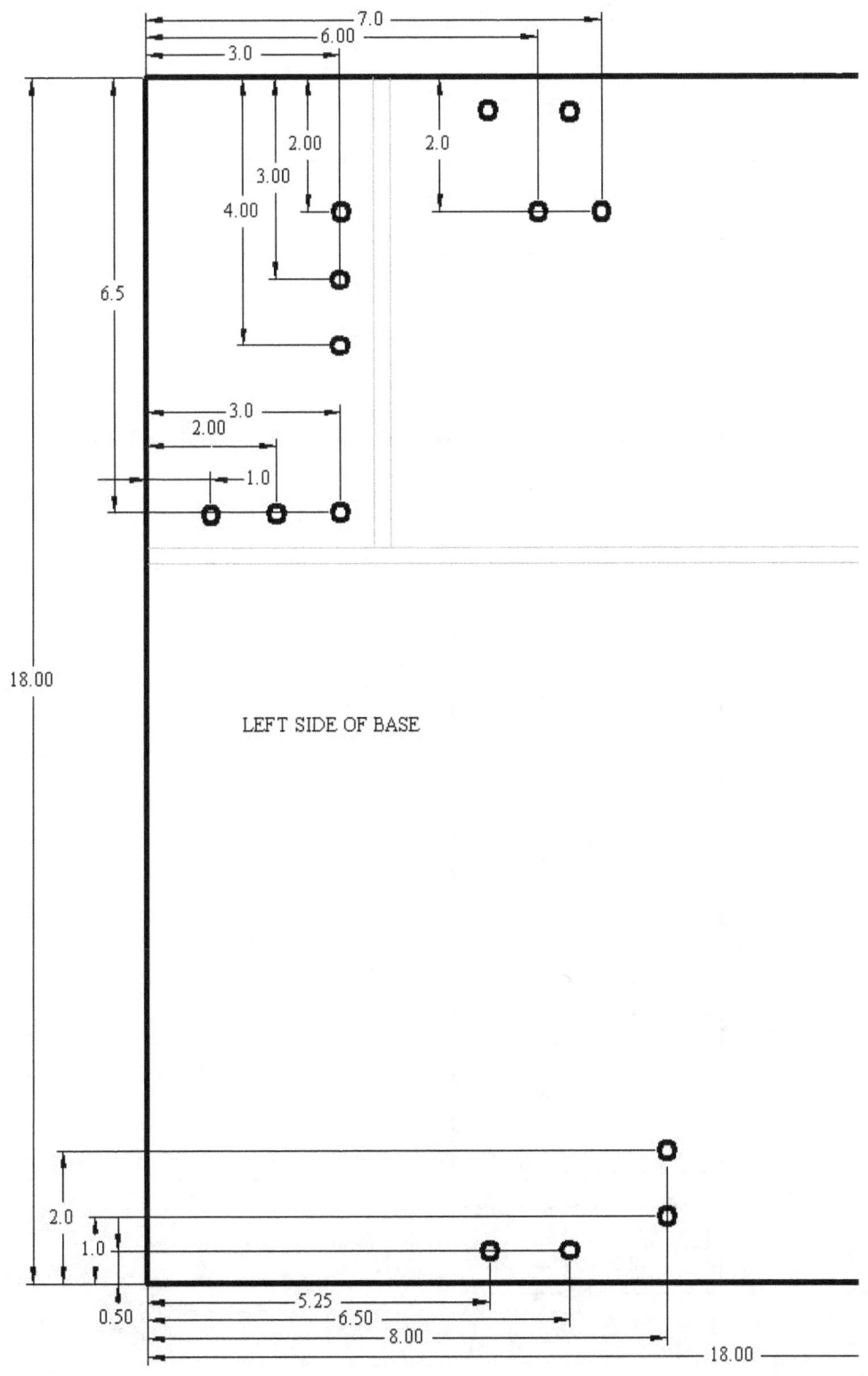

LEFT SIDE OF BASE

Mendel 90 Gantry Left Side, the right side is completely symmetrical.

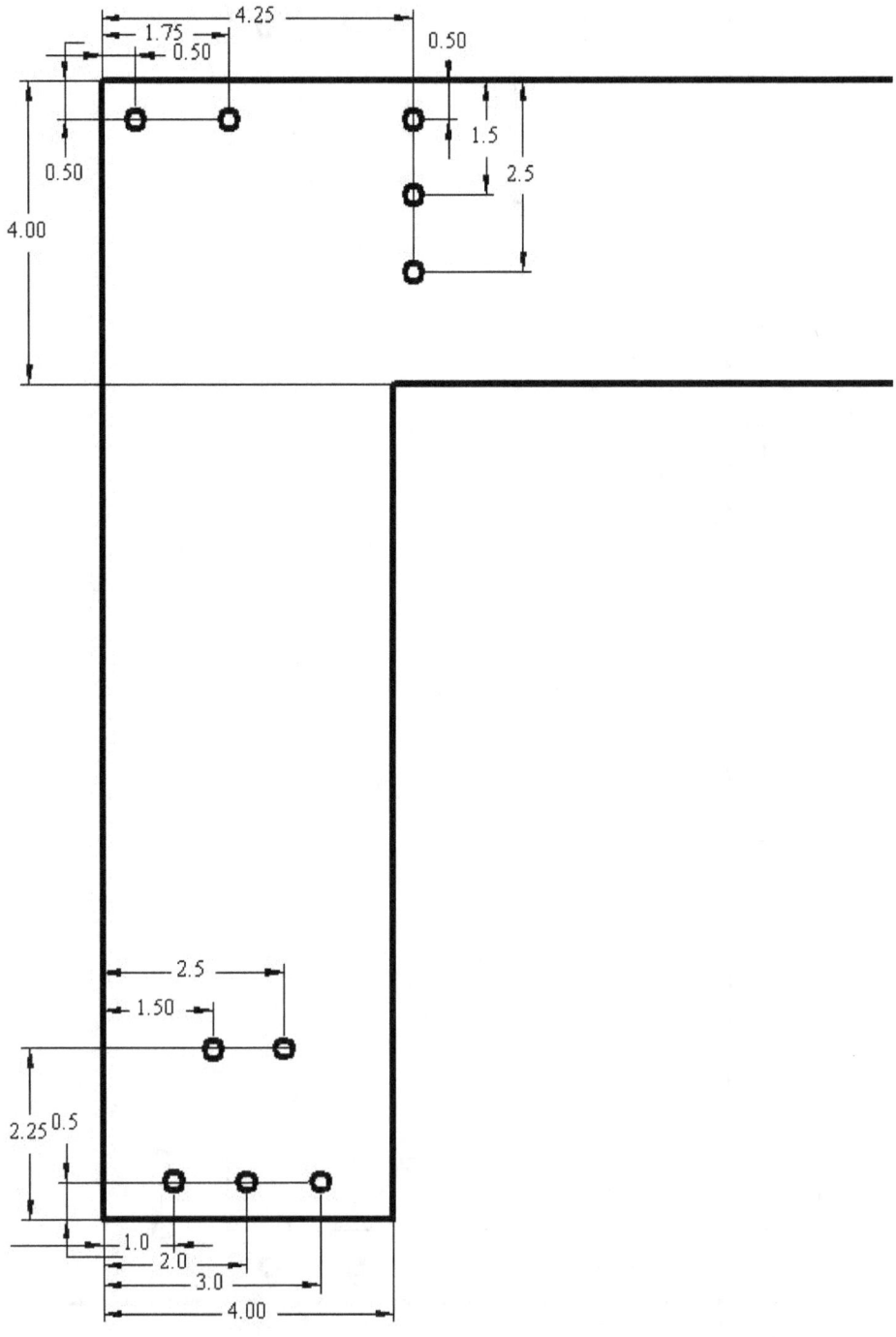

Sorry but the Prusa i3 gantry design is in metric unless you convert it to inches. You will need to add some holes for fastening the angled braces. The braces are mounted almost flush with the inside edges. The braces are the same as for the Mendel 90 but they are only 14 5/8 inches tall for the Prusa i3. The gantry is symmetrical and the two holes at the top are the same horizontal spacing as the two holes at the bottom.

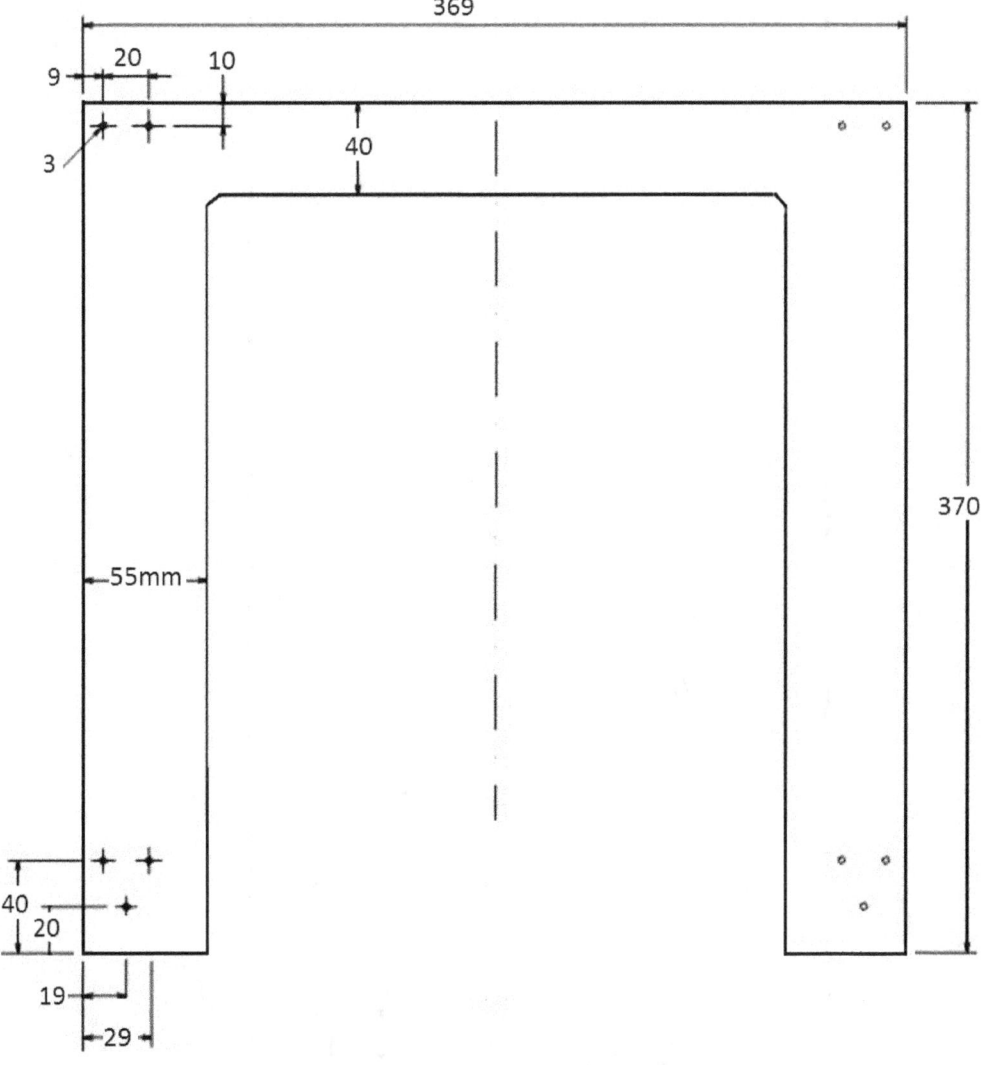

Some conversions to inches are: 370mm / 25.4 is 14.57 inches, 55mm / 25.4 is 2.16 inches, 40mm /25.4 is 1.57 inches, etc.

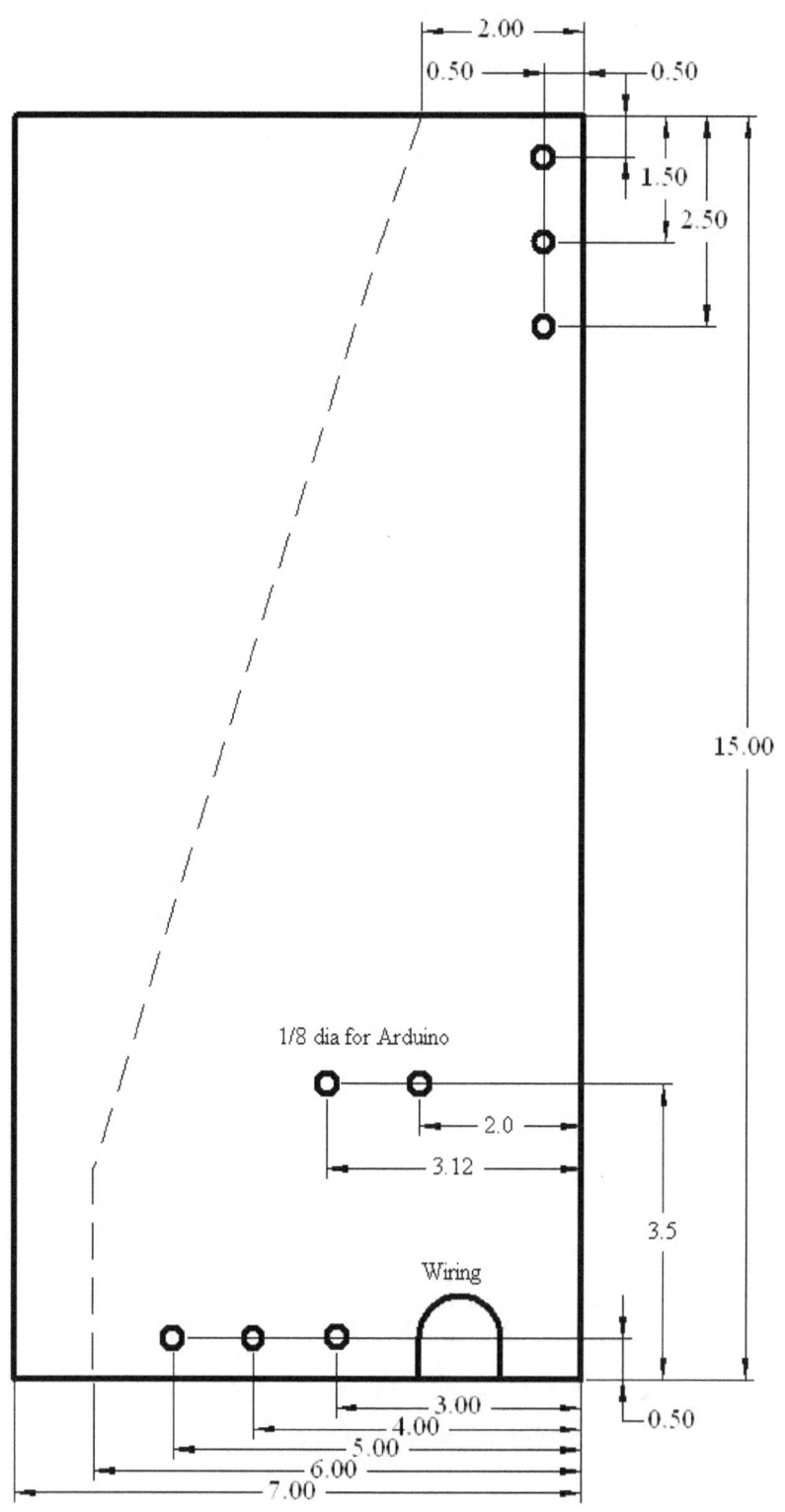

1/8 dia for Arduino

Wiring

2.00
0.50
0.50
1.50
2.50
15.00
2.0
3.12
3.5
0.50
3.00
4.00
5.00
6.00
7.00

The gantry braces can be cut on the dotted line to save on Plexiglas. The gantry brace on the right (or left) side should have some holes in it for mounting the Arduino Mega as well as a large notch at the bottom for the many wires to pass through.

To get better clearance for the outer holes, I came up with a design for making an oversized Y axis carriage. Basically this design adds .25 inches to all four sides thus making it nine inches by nine inches. Most of the holes are 6-32 clearance holes except for the corner holes; they are 4-40 clearance holes. This drawing is for 6.25" rail spacing. The Prusa i3 uses 6.75 inch rail spacing, adjust the numbers accordingly.

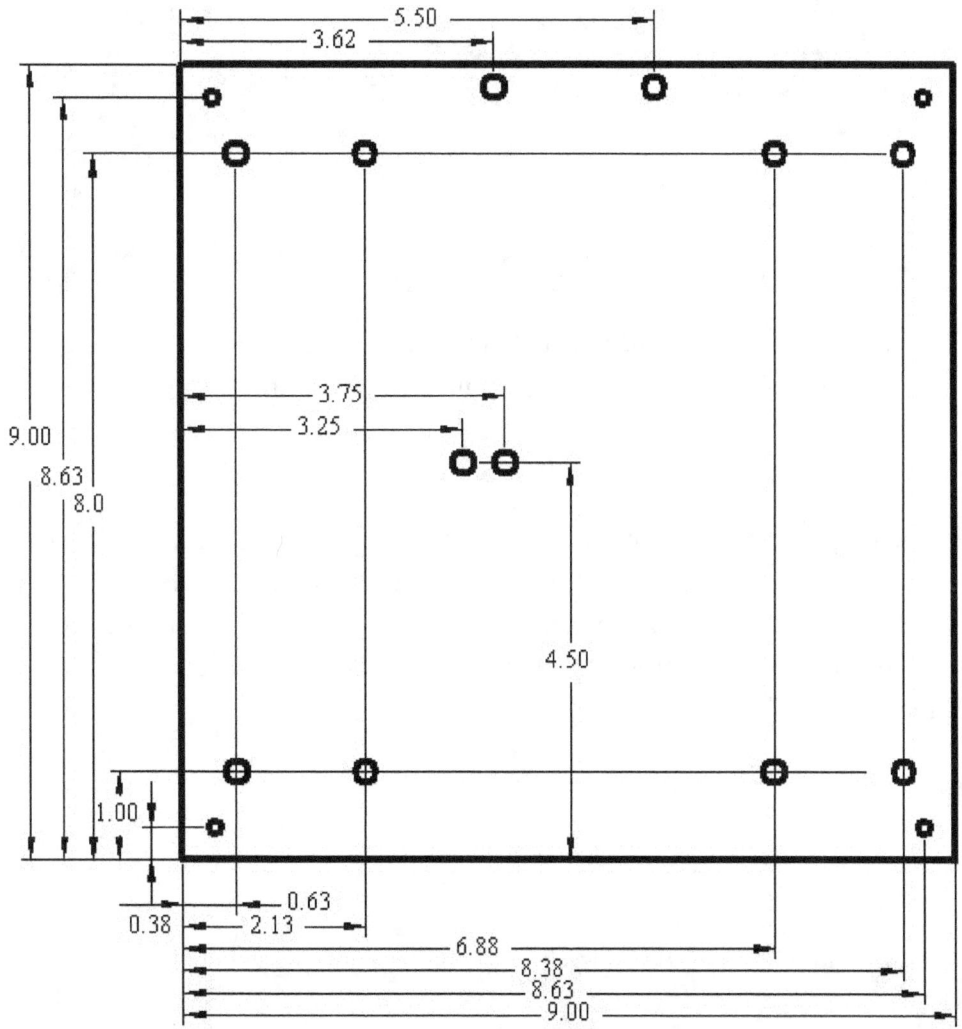

Next there is a picture of the drilled and almost assembled oversized Y platform. The bearing mounts are 1.5 inches across so they will fit the 1/2 inch pipe holders used for 8 mm linear bearings or, in my case, the 3D printed 10 mm linear bearing holders.

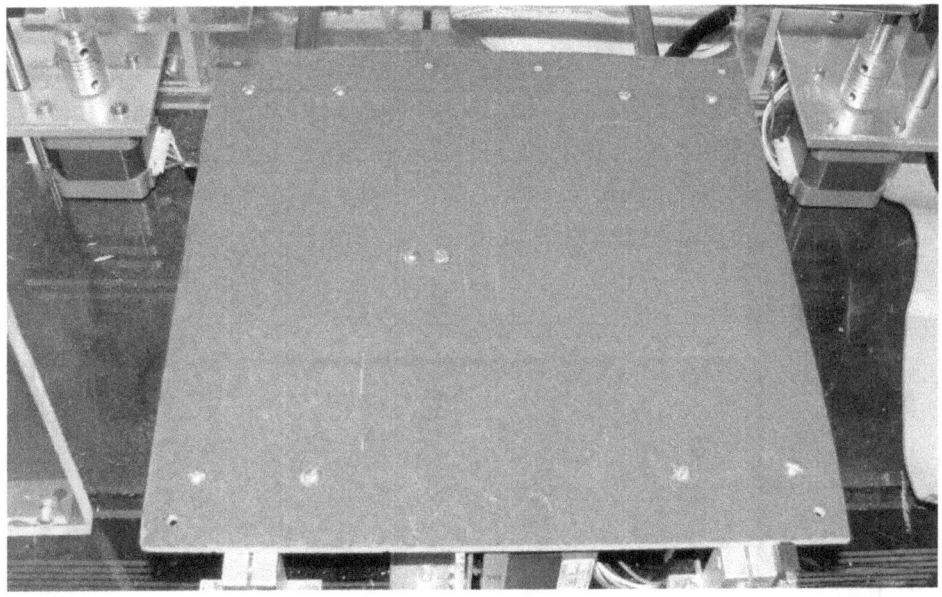

Coming up next is the mechanical drawing of the left side Z axis motor mount and guide rod support. The right side is a mirror image of the left side. The position of the guide rod hole is off just a little. I had to file it a little so that the hole is a tiny bit closer to the stepper motor.

The large holes are for the stepper motor and guide rods, they are 5/16 of an inch or 8 mm in diameter. The mounting holes are 8-32 clearance holes or about 5/32 of an inch in diameter. The stepper motor holes are for the 3mm metric screws that are needed to be used to mount the stepper motors.

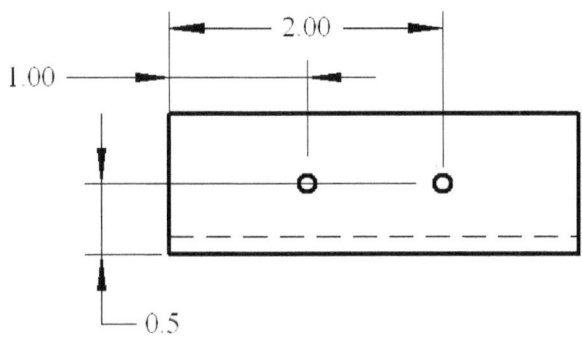

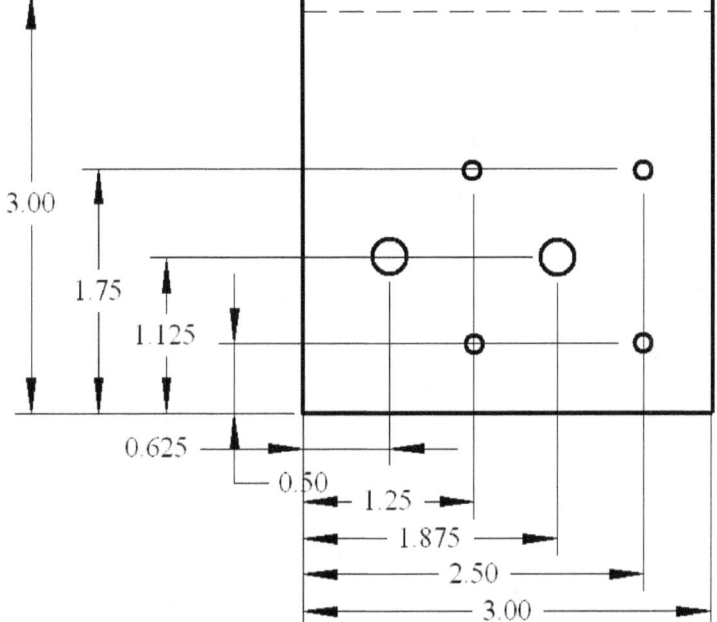

Up next is the mechanical drawing for the Geeetech right motor mount. I made 4 motor mounts, two for the top and two for the bottom. You will need one inch by 2.5 inch L shaped .125 inch thick aluminum. I did not have that size so I used one inch by three inch "C" channel aluminum instead. Even after careful measuring I had to file the holes slightly to reduce the spacing between the stepper motor and the guide rails. My careful measurements were for 21 mm motor to rail spacing.

The mounting holes are .5 and 1.28 inches from the right side and on the center line of the 1 inch side. They are for 6-32 screws to be compatible with the Prusa i3 frame.

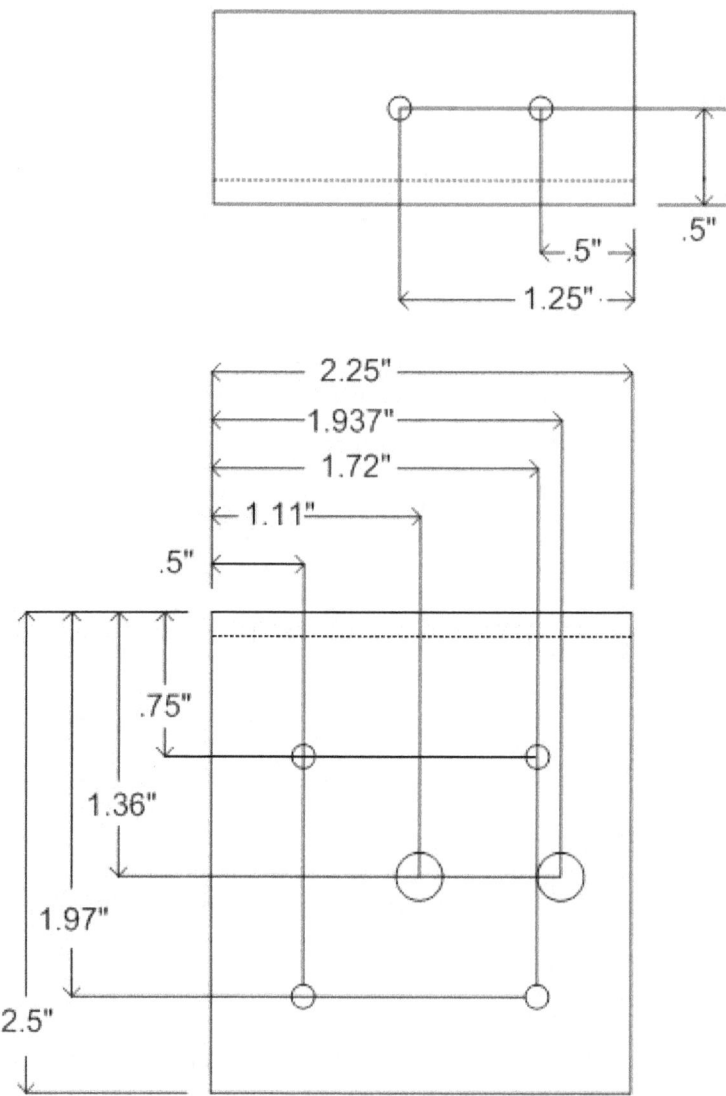

The upcoming picture shows what the cut and drilled motor mounts should look like. I used 6-32 nuts for spacers in each corner to space the stepper motor down so there was no need to drill the large 7/8 inch holes for the motor bearings that are on the front of the motors.

The back motor mounts in the picture are for the Prusa i3, the middle motor mounts are for the Geeetech i3, and the front motor mounts are for the Mendel 90 and were modified for two different rail spacing.

This next drawing is of the Y axis motor mount. The height shown of two inches is not correct; it actually needs to be about 1.95 inches tall. You can also buy a standard NEMA 17 motor mount and use that if you want. In that case you might need to file it slightly so that it does not scrape on the Y axis as it passes over it.

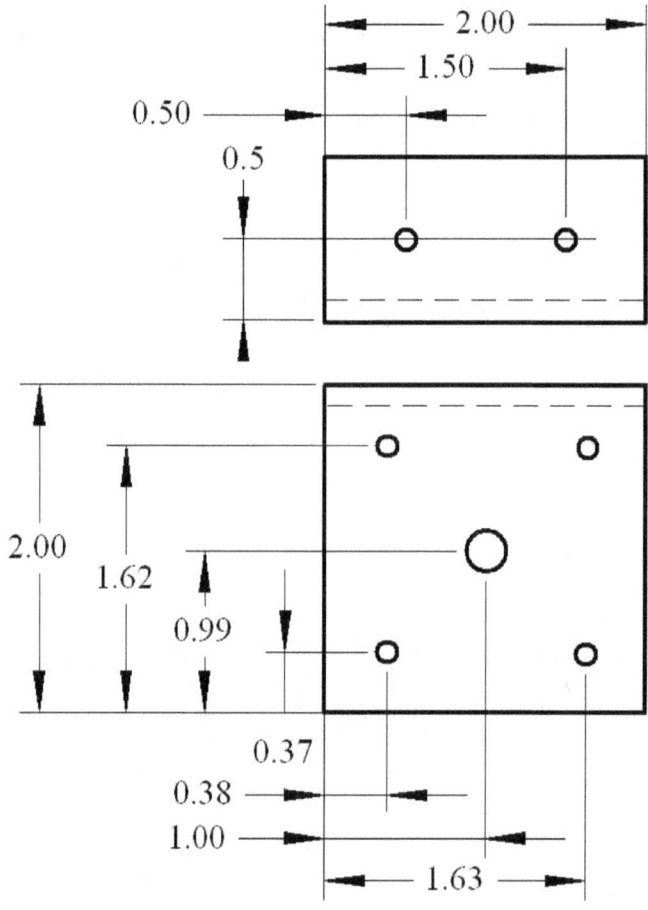

Here is a picture of the completed Y axis motor mount.

You can also make your own X axis platform if you want to. The platform can be made out of a piece of Plexiglas or aluminum about three inches by four inches in size. The large center hole should be 1.5 inches in diameter although it is only shown as one inch in the drawing.

Making your own X axis platform is optional, you can just buy one on eBay that is already made and the bearings just snap into it. What I do not like about the standard design is the use of only three bearings instead of four bearings. Also the standard design limits how you mount the extruder. The only mounting holes are on the right and left side of the platform.

I improved this design slightly. The bearing straps are 3/8 of an inch from the edge. This change gives more clearance to the hot end. Also I countersunk the holes for the bearing holders and used flat head 6-32 screws so the extruder can mount down flush to the surface of the carriage. This X axis platform design only works for the Mendel 90 arrangement.

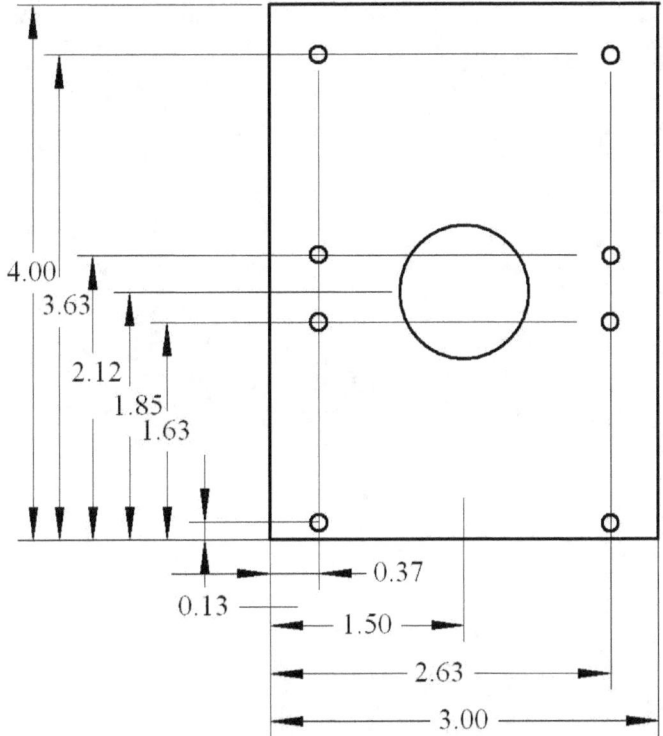

Another option is to buy a metal X axis metal carriage as can be seen in this picture. This picture also shows a Prusa i3 wood frame with metal motor mounts. The wood frame looks better with three or four coats of gloss black spray paint.

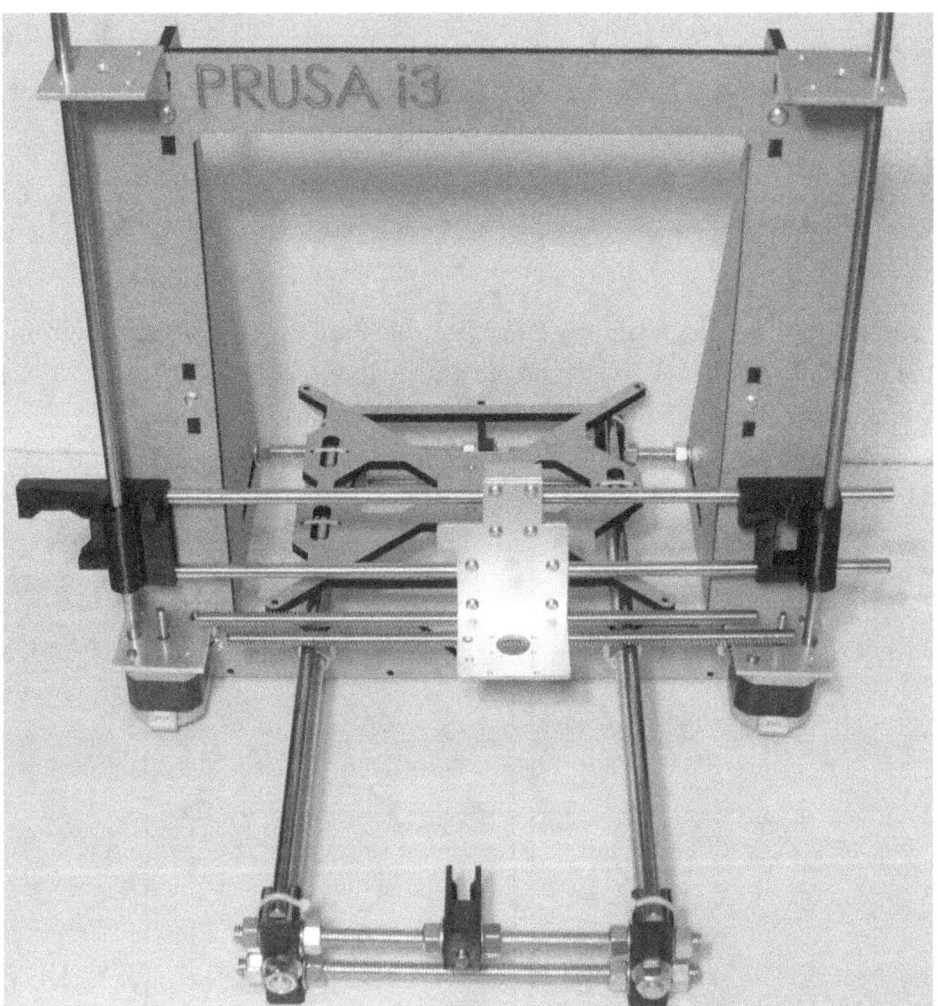

Chapter 5
Mechanical Assembly

I decided to use 3/4 inch 8-32 screws and nuts for most of the assembly as 6-32 was too small and allowed for too much play in the parts. You will need about 80 of them!

Here is a brief list of needed tools;

  Drill and drill bits (Drill press is better)
  Good Jig Saw or table saw
  Square
  Caliper
  Screwdrivers
  Allen wrenches
  Files (Need a small round file)
  Ruler with both metric and inches
  Soldering Iron and solder

Here are some pictures of a few of the tools that I use.

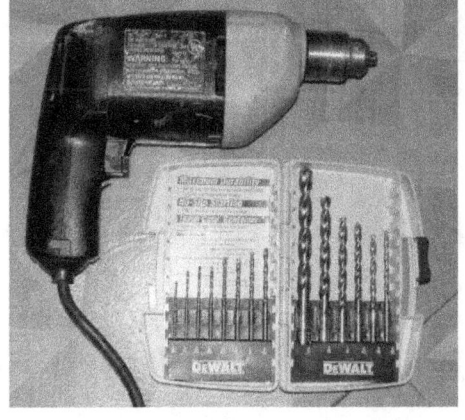 

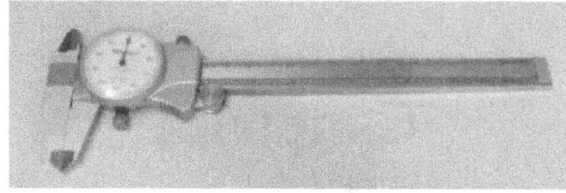

Assemble the base first. Add the metal braces then add the gantry sides, and then add the gantry braces.

Add the 10mm rod holders, rods and bearings. The 10mm rod holders do not tighten enough to hold the rods in place. You might need to add a layer of clear tape to fix that problem. This is what it should look like.

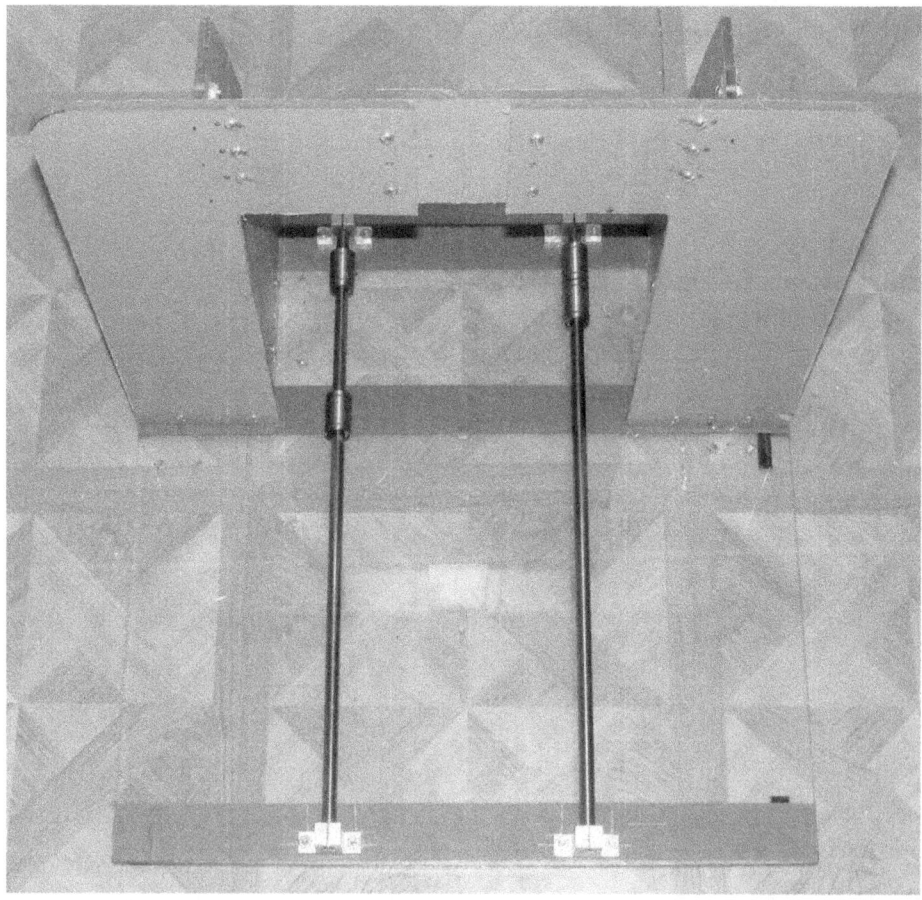

The 10mm rod holders need to be spaced up at least 3/4 an inch. If not then there is not enough clearance for the motor to fit underneath the platform. The screws will need to be almost 2 inches long to fit through those spacers. You can see by the extra holes that it took me a few tries to get the hole spacing correct.

If you are building the base that is made out of 8mm or 5/16inch threaded rod, then build the short ends first. You will need one 36 inch threaded rod cut in half and four 12 inch threaded rods; three of them need to be cut to 9 inches. Center the bearing and motor mounts on the ends.

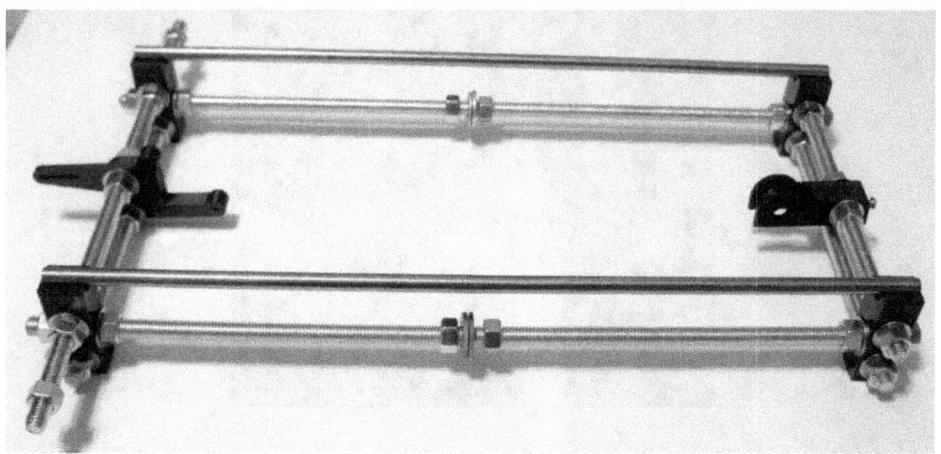

The length matches the longest 8mm rails. The width is 6.75 inches but it is best use the frame if available to fine tune the width.

After the motor mounts are added you can add the 8mm rod holders but do not tighten the rods. That way the rods can be easily removed to install or remove the X axis assembly to work on it. For the Mendel 90 the rod holders will need spacers that are a little over one inch tall. I used threaded spacers with two washers to get the correct amount of spacing.

With the printed Prusa i3 motor mounts and guide rods installed your printer should look something like this next picture.

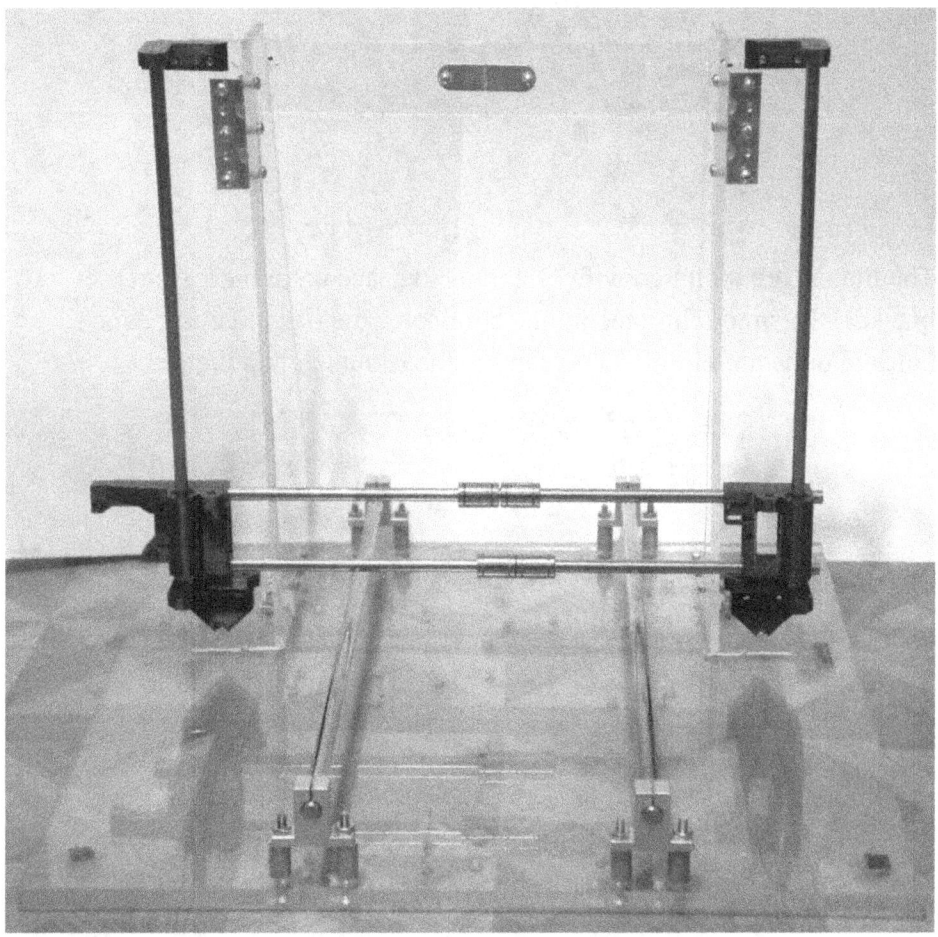

This next picture is what the Y axis belt idler looks like. It is a 1/4 inch bolt that passes through two 1/4 inch flanged bearings then a nut then a two inch "L" Bracket from the local hardware store and then another nut. The top of the bracket needs to be either bent over or cut off because there is just a little less than two inches clearance. The outside diameter of the 1/4 inch bearings is about 5/8 of an inch before the flange.

The motors are mounted with 6-32 nuts to space them back from the bracket. This mounting technique eliminates the need to cut 7/8 inch holes into the aluminum. You can see those nuts in this picture.

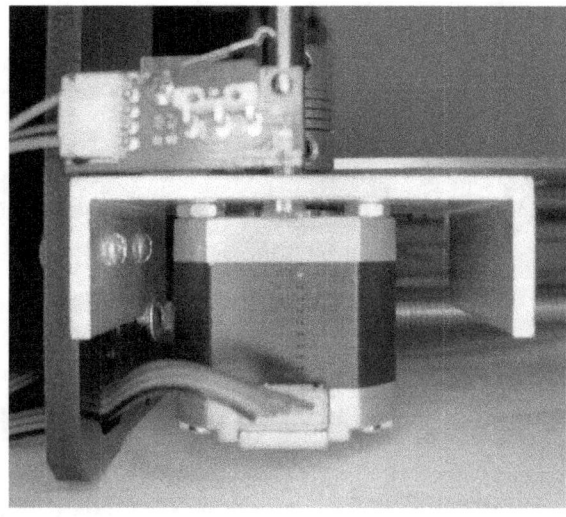

To assemble the X axis install the stepper motor with the belt gear installed backwards on its shaft. I did not do that at first and had to take it back apart. Then strap the 8mm linear bearings onto the X axis ends with plastic wire ties. In the Prusa i3 the bearings slide in place. Insert the bearings on the X axis carriage and put the carriage on the two 12 inch long 8 mm rods and then insert the rods into the X axis ends. The i3 uses 15.5 inch rods for the X axis.

Install the nuts and thread the lead screws through the nuts. Then place the X axis assembly onto the Z axis stepper motors. Then carefully insert the Z axis guide rods down through the X axis end bearings. At this point your machine should look something like this next picture (back view).

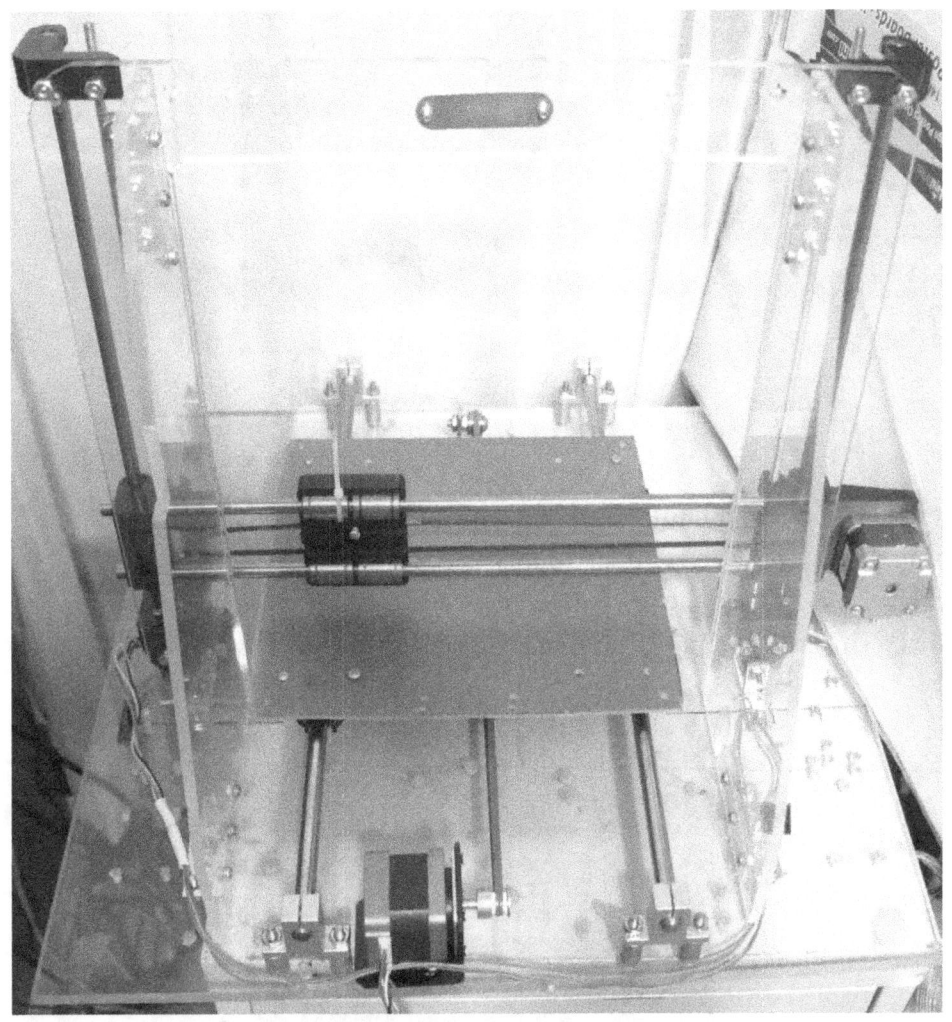

You can optionally build a heat shield that goes between the Y platform and the heat bed. This shield protects the Y platform from the heat and insulates the bottom of the heat bed so it does not waste as much heat from the bottom side.

The heat shield is made out of one layer of corrugated cardboard. It is cut into an 8.5 inch by 8.5 inch square. Then the corners are cut out by removing 1/4 inch squares from each corner. Then the piece of cardboard is taped over with duct tape. Be sure to cover the open ends so that air cannot flow out of the ends of the cardboard.

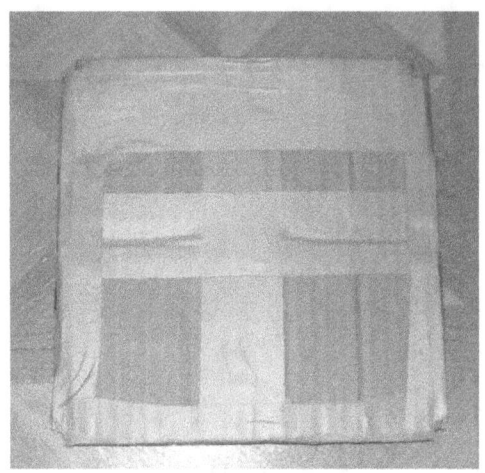

Finally a layer of aluminum reflective tape is put over the top side of the shield. You only need about 26 inches of reflective tape if it is three inches wide. Then the assembled heat shield is placed underneath the heat bed. This next picture is what the completed heat shield looks like.

I have mentioned that you can make your own X axis carriage. It will require the use of four LM8UU bearings, four 1/2 inch plastic pipe holders and a piece of Plexiglas about three inches by four inches in size. This is what that looks like. This is for the Mendel design only.

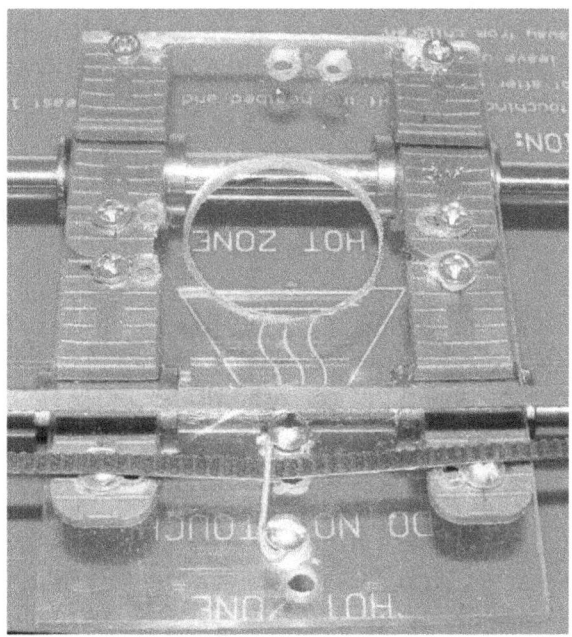

The front holes are a little crooked because they were preexisting. The back two are for mounting the extruder and two of the front ones are used for the belt clamp. The center hole was 1.25 inches and that was not quite big enough to fit my extruder hot end down through it. The pipe straps need about 1/8 of an inch to be cut of their back sides and even then they tend to scrape up against the back walls of the gantry.

Some of the **problems and solutions** that I have encountered in the mechanical assembly are listed below.

The stepper motors came with gears on them. I used a gear puller to remove the gears. When I connected the motors up to the lead screws I discovered that their shafts were bent so much that the top of the lead screws rotated in about a one inch circle. I had to get new stepper motors.

I was stuck on the lead screws for the Z axis. The top holes for the nuts are .55 inches across and they are supposed to be for 8mm nuts. But 8mm nuts are .51 inches and 5/16 nuts are .5 inches. Both nuts rotate freely in the hole as the hole is for a nut that is .55 inches across. It turns out that the nut holes on the top are .55 inches but the nut holes on the bottom are .51 inches! So I was just looking at the wrong nut holes.

When I assembled the Y axis carriage and slid it towards the back it returned to the front on its own! That was because the rod spacing varies by 1/8 of an inch causing the bearings to bind. I enlarged the holes on the back side and adjusted it so the platform moves smoothly now.

The 8mm bearings I bought are simply terrible. They do not move smoothly and they tend to lose some of their balls every time I remove them from the rods. My solution was to hand pick the bearings that are used for the ones that move the smoothest.

I used 10mm rods (from scanners) and bearing for the Y axis. They work so much better than the 8 mm bearings. However, their outside diameter does not fit pipe holders. I ended up using printed bearing holders.

While testing the Y axis I realized that the gantry is at the wrong position. When I move the Y carriage back the print head will not reach the front edge. I had to unbolt the gantry and move it forward one inch to resolve the problem. It was six inches from the back edge but it is now seven inches from the back edge.

The Y axis bearing holders hits the stepper motor that is mounted under it. That is because I mounted the stepper motor all the way to the right. It had to be moved at least an inch to the left. It was moved so that the Y platform belt can be run in the exact center between the rails.

I cannot put nuts on the Y axis hot plate screws. The bearing holders are too close to the front and back edge of the Y platform so there is not enough room to put the nuts on the hot plate screws. I moved the Y axis bearing holders from 1/2 to 1 inch from the front and back edges of the Y platform. This also added about 1 inch to the range of the Y platform.

.

Chapter 6
Electronics Assembly

I started by making six pin to four pin cable adapters for the three stepper motors in the base (Y and Z axis). They can then be plugged in and even tested out. You may need to make several four to six wire adapters if you have to make your own cables.

The Y and Z stepper motors are wired up directly via six pin to four pin adapter cables. Most stepper motors have six wires but we only use four wires. The reason is that the extra two wires are the center taps of the coils. These center taps are used for stepper motor setups where the center taps go to power and each of the four other wires goes to a driver transistor. We are not doing that here. Here is the translation from six wires to four wire operation.

6 wire Pin	4 wire Pin
1	1
2 (Center tap)	
3	2
4	3
5 (Center tap)	
6	4

If you should accidentally reverse the order of the four pins, then the motor will just spin in the wrong direction, but no harm is done.

The electronics wiring can use two or three ribbon cables. Each cable is about 20 to 22 inches long. They can be salvaged from any old personal computers. Cables with the older .05 inch spacing work the best, anything smaller than that is really hard to solder.

You will need 9 or 15 wires (with a 9 or 15 pin plug and socket) for the extruder if you want a plug in extruder. You will need 25 wires (with a 25 pin plug and socket) if you want a plug in X axis.

This can be a continuation of the 15 pin cable to save splicing cables. You will need 26 wires for a ribbon cable for the heat bed. It is usually hardwired.

These listings are for the 9 or 15 Pin connectors to the extruder. It is in numerical order. The 15 pin connector has two rows one of seven and one of six pins. The 9 pin has rows of five and four pins.

Pin	Signal
1	Thermistor
2	NC
3	Heater +12V
4	Heater –
5	Heater –
6	Extruder Stepper 1
7	Extruder Stepper 3
8	NC
9	Thermistor Ground
10	Heater +12V
11	Heater +12V
12	Heater –
13	Fan –
14	Extruder Stepper 2
15	Extruder Stepper 3

Pin	Signal
1	Fan +12V
2	Fan Ground
3	NC
4	Thermistor
5	Thermistor
6	Extruder Stepper 1
7	Extruder Stepper 2
8	Extruder Stepper 3
9	Extruder Stepper 4

Use a separate connector for Heater + and -.

This is the pinout for the optional 25 pin connector in numerical order. This connector has two rows, one of 13 pins and one of 12 pins.

Pin	Signal
1	X Limit
2	X Limit Ground
3	Thermistor
4	Thermistor
5	NC
6	Heater +12V
7	Heater +12V
8	Heater +12V
9	Heater –
10	Heater –
11	Heater –
12	Fan –
13	Extruder Stepper 1
14	Extruder Stepper 2
15	Extruder Stepper 3
16	Extruder Stepper 4
17	X stepper 1
18	X stepper 2
19	X stepper 3
20	X stepper 4
21-25	NC

The arrangement might seem strange at first but the cable is logical in its wiring arrangement. The stepper motor wires are kept together to reduce interference, and the thermistor wires are kept away from the stepper wires.

The heat bed takes a 26 conductor ribbon cable this is how it is wired up.

Pins	Signal
1-12	Heatbed Power
13	Thermistor
14	Thermistor
15-26	Heatbed ground

There are schematics of the Ramps shield available on the Internet but they generally have mistakes, omissions, and they make it look way too complicated. So I have created my own schematics to make working with the shield much easier.

Next up is the schematic diagram of the end stop connections. In actuality the 5V pin is on the left, ground in the middle and signal on the right side.

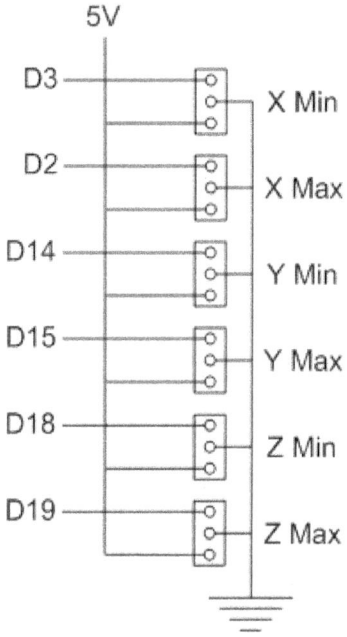

Here is the schematic of the thermistor input circuits.

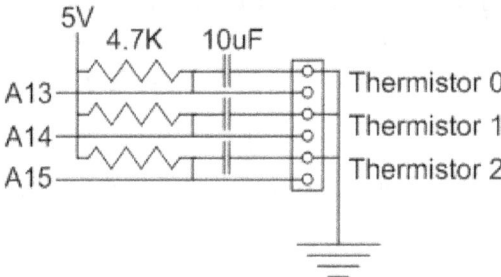

Up next is the schematic of the heater driver circuit of the Ramps shield.

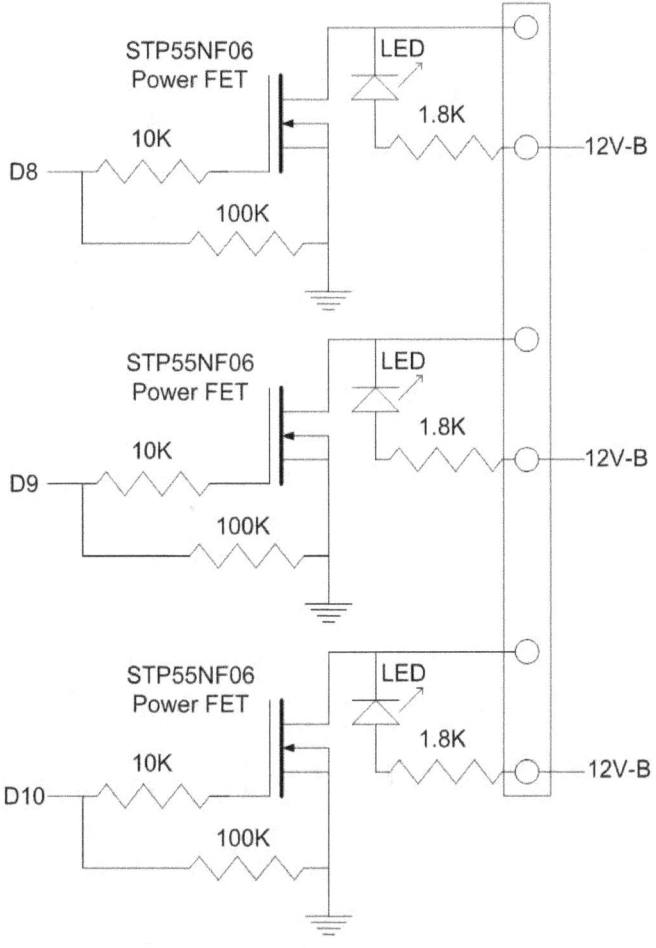

Up next is the internal wiring diagram of the Ramps stepper motor drivers. The schematic is missing pull up resistors that are on the "enable" inputs. On the other hand all of the pin connections to the Arduino Mega are listed as well as the purpose of each signal. All of the three jumpers to 5V on the inputs to the A4988 step sticks are installed by default and are left that way.

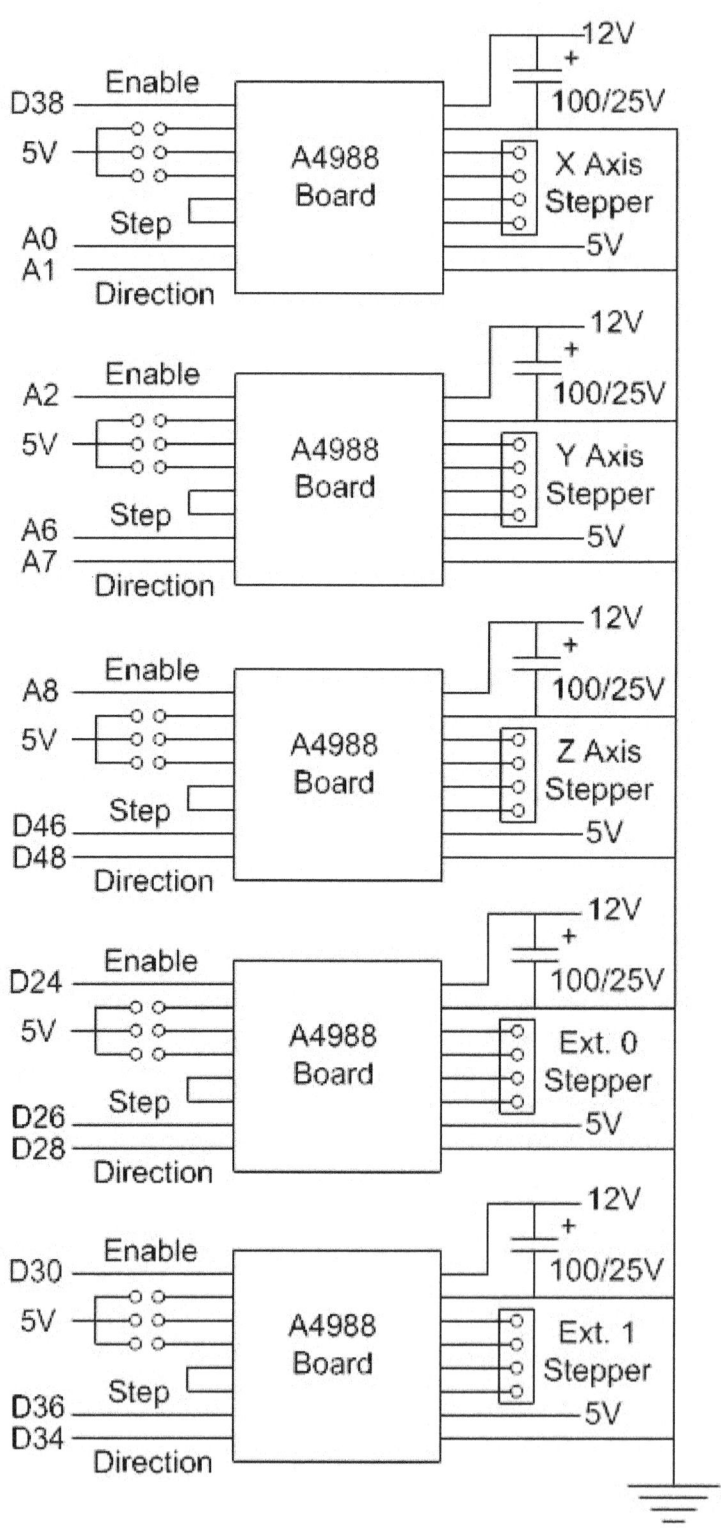

Here is the wiring diagram for the Arduino Mega Ramps 1.4 shield. This diagram is also commonly available on the Internet, but I have rotated it 90 degrees and converted it to gray scale so it appears like it will when it is actually being wired up. I also fixed the wiring to the end stop switches so that they are now correct. The wires go to the outside terminals or the "NC" terminals of the switches. If you are using three wire switches the red wire goes to the left. "Hotend" is the hot end of the extruder. "Heatbed" is the heater circuit board that is on the top of the Y axis.

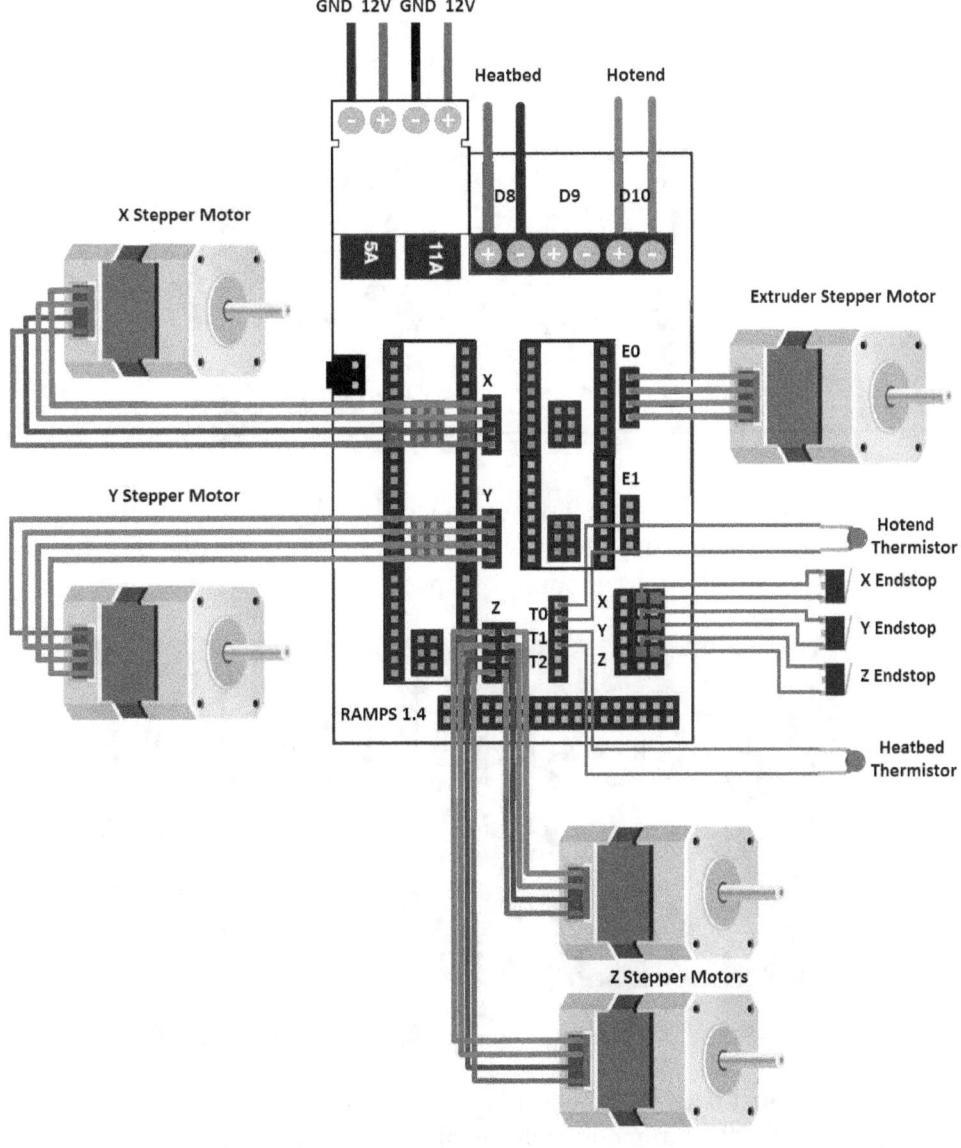

This next picture is of the extruder's 15 pin plug once it was wired up. I had to make a small circuit board with the 15 pin plug on it in order to space the plug up about 1/4 an inch above the top of the stepper motor. Otherwise the ribbon cable would scrape on the belt. The connectors for the fan and stepper motor have three and four pin plugs so that the fan and stepper motor can easily be unplugged for working on the extruder.

This next picture shows the 9 pin extruder plug in use. It is much smaller, easier to make and to use. However it requires a separate plug for the heater. The heater draws about 3-4 amps so it is best to use a connector that can deliver that current.

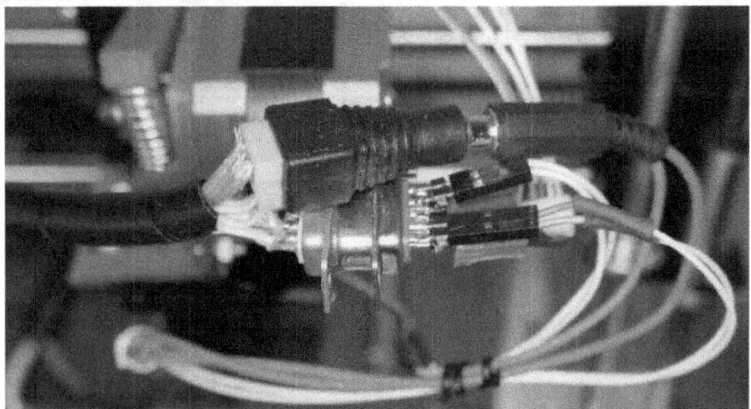

Coming up next is a picture comparing the 9 pin and the 15 pin adapter. As you can see the 9 pin connector is much easier to wire up as no circuit board is needed. Basically you just solder two 2 pin and one 4 pin connectors to the back side.

 Limit switches are used to detect the edges of the printable area as well as when the extruder contacts the glass surface Z=0). The X limit switch is located at the right side of the X axis. You can make an X axis limit switch out of a piece of fiberglass circuit board and a micro-switch. It is easier to buy limit switches with the board and wiring all ready to install.

The X limit switch bracket mounts on top of the right X axis end assembly. A screw or two go through the board and the opening in the X axis end. It is easier to do when the belt is not installed. The Geeetech switch is harder to mount. The space is very limited. It is mounted on the top of the left side of the X axis. I cut the board off right after the switch and then soldered the wires directly to the switch. It is then mounted using one screw and an insulating spacer as is seen in this picture. If done correctly the LED will still work!

The Y axis limit switch is mounted on the side of one of the Y axis rail holders. I used a metal L bracket to hold the switch in place. It was at the front for the Mendel 90, but at the back for the Geeetech i3 design.

The Z axis switch needs to be mounted so that it can vary its height above one of the Z axis stepper motor mounts. There is a printed part that does just that. Use the stepper motor screws to hold the bracket in place. You can do some adjusting by bending the metal lever that is part of the switch up or down.

The Geeetech Z axis switch is stationary but a long screw mounted on the X axis end can adjust when the switch will trigger. A simple L bracket can be used to mount the switch.

Wiring the heatbed can be a little tricky. The ribbon cable is not as wide as the circuit board traces are. It takes a double bend to make them wider. Then right below where the ribbon cable is soldered on I added a metal strap to fasten it to the Y axis carriage. The cable then loops to the back of the machine and then back up to a point across from the wiring hole in the gantry support. There the ribbon cable is broken down into groups of four conductors except for the two conductors for the thermocouple. That way they can pass through the small hole in the gantry brace.

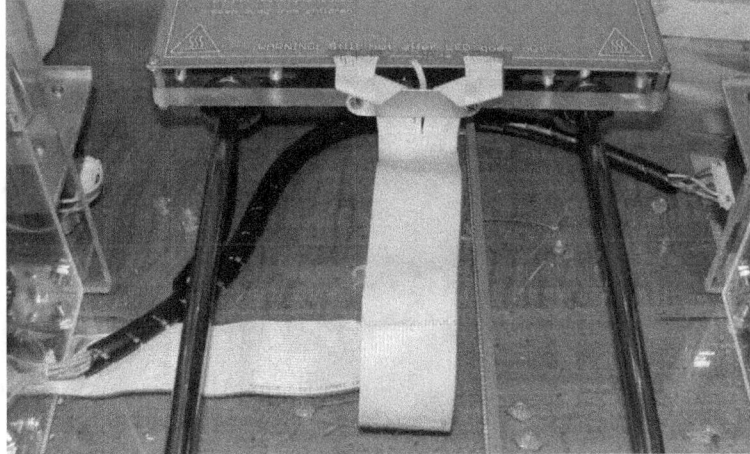

You can use some 14 gauge electrical wire or speaker wire for the heat bed instead. I also used some spiral wrap to help protect it from catching on the surrounding parts. This method will require some room behind the printer for the wire to bend back into.

You will need a 12 volt 12 Amp power supply to run the heatbed. Some people just use an old computer power supply. I used a 12 volt power supply that I had lying around. I added a 120 volt power plug adapter hanging off the end of the power supply. Then I used some large speaker wire to connect it up to the 3D printer. The wire from the power supply to the printer should be at least 14 gauge stranded wire. If it is not big enough it will drop a volt or two and get warm. I later upgraded to some used 110 volt 14 gauge electrical wires going from the power supply to the RAMPS to get more power to the heat bed. There should be an insulated cover over the connecting screws on the power supply in this picture.

This next picture is what the RAMPS shield wiring looks like. It took some rearranging to make the wiring look this neat. You can see the black spiral wrap around the wires coming from the heat bed on the left side as well as around the wires coming down form the X axis on the right side.

The next two pictures shows what the 3d printer looks like once it is all wired up in two different arrangements. The top picture is the home made gantry with the Prusa i3 printed parts. The bottom picture is the home made gantry with the Geeetech X axis and homemade motor mounts. Not visible in the pictures is the 12 volt 5 amp AC adapter that provides power to the stepper motors.

It took me about two hours to do the mechanical conversion from one set up to the other. Changing the software to work with the different arrangement is a far different matter. There will be more on the software in the next chapter.

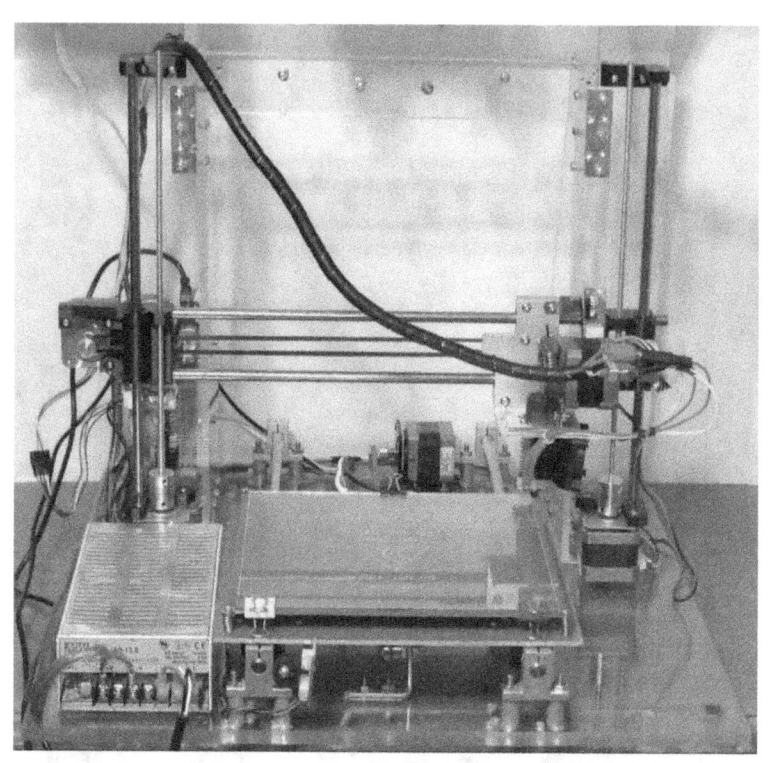

Chapter 7
Software Setup

We will be using a number of programs to get this 3D printer working. First we will be using "Teacup" or "Marlin" for the "firmware" that goes inside of the Arduino Mega. To send the firmware to the Arduino Mega we will use Arduino 1.0.4. On the Computer we will be running "Printrun" to send machine codes to the 3D printer, with "Slic3r" to take 3D objects and "slice" them into "layers" to create the needed machine code. The PC and Arduino are connected to each other via a standard USB cable.

To install the Arduino interface drivers first download Arduino 1.0.4 windows.zip or a newer version. Then decompress the files into a known location. I like to use the root directory of the C: drive so that the files are easy to find.

The decompressed files create a directory called "Arduino-1.0.4" with all of the needed files. You can now plug in the Arduino Mega with a standard USB cable. Windows should come up with the "Install new hardware" wizard. You will need to tell it to "Install from a specific location", and then direct it to the "Arduino-1.0.4\drivers" directory. Everything should now install and Windows will then recognize the Arduino Mega.

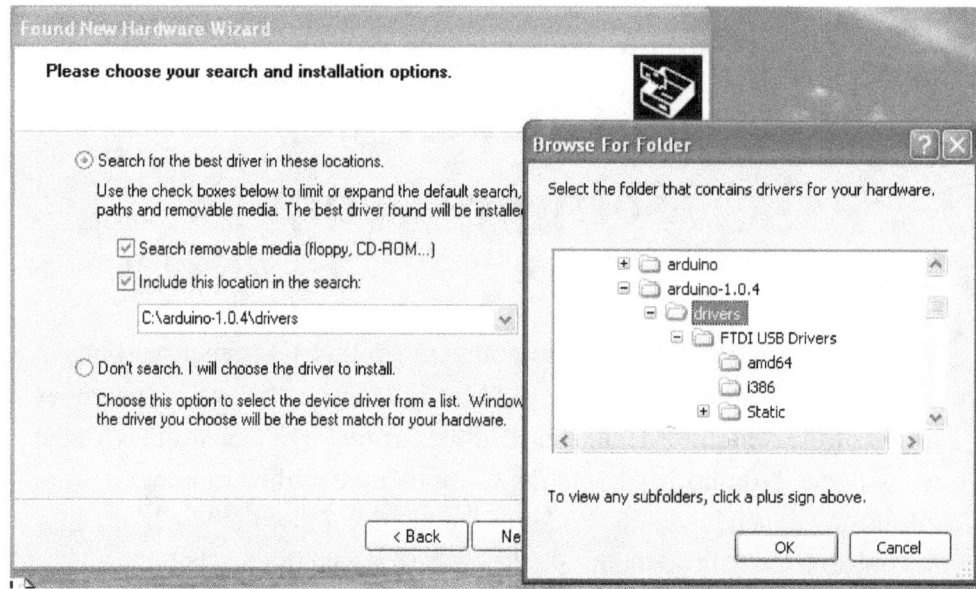

Also in the Arduino-1.0.4 directory there is a program called "Arduino.exe". It is the program that you will use to communicate with and upload your firmware, and revisions to the Arduino Mega. It is a good idea to create a shortcut to it on your computers desktop so that it is easy to find.

There is also a subdirectory of Arduino-1.0.4 that is called "reference". In that subdirectory there is a file called "index.html". When you double click on that file a reference library should come up in your favorite browser. This library can be very helpful in debugging your code. It can also be accessed within the Arduino.exe program by selecting "help" and "reference".

Next you will want to set up "Teacup" on your Arduino Mega. Download and then Unzip "Teacup.zip". I put it right in my Arduino software folder so it is easy to find once it is unziped.

Name	Type	Modified
.gitignore	GITIGNORE File	8/31/2014 10:32 ...

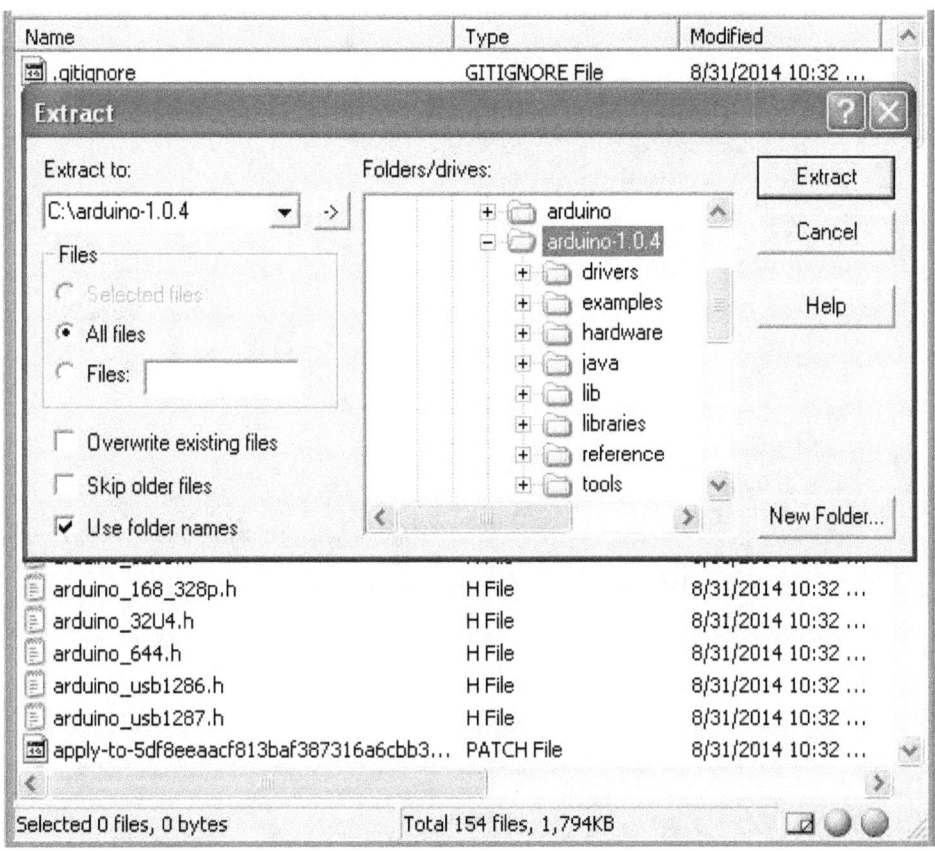

arduino_168_328p.h	H File	8/31/2014 10:32 ...
arduino_32U4.h	H File	8/31/2014 10:32 ...
arduino_644.h	H File	8/31/2014 10:32 ...
arduino_usb1286.h	H File	8/31/2014 10:32 ...
arduino_usb1287.h	H File	8/31/2014 10:32 ...
apply-to-5df8eeaacf813baf387316a6cbb3...	PATCH File	8/31/2014 10:32 ...

Selected 0 files, 0 bytes   Total 154 files, 1,794KB

Rename teacup so it is just "Teacup_Firmware" remove anything else that is in its name. You can do that by right clicking on the file name and selecting "Rename" This can be seen in the next picture. The same renaming technique will be used for the next few changes.

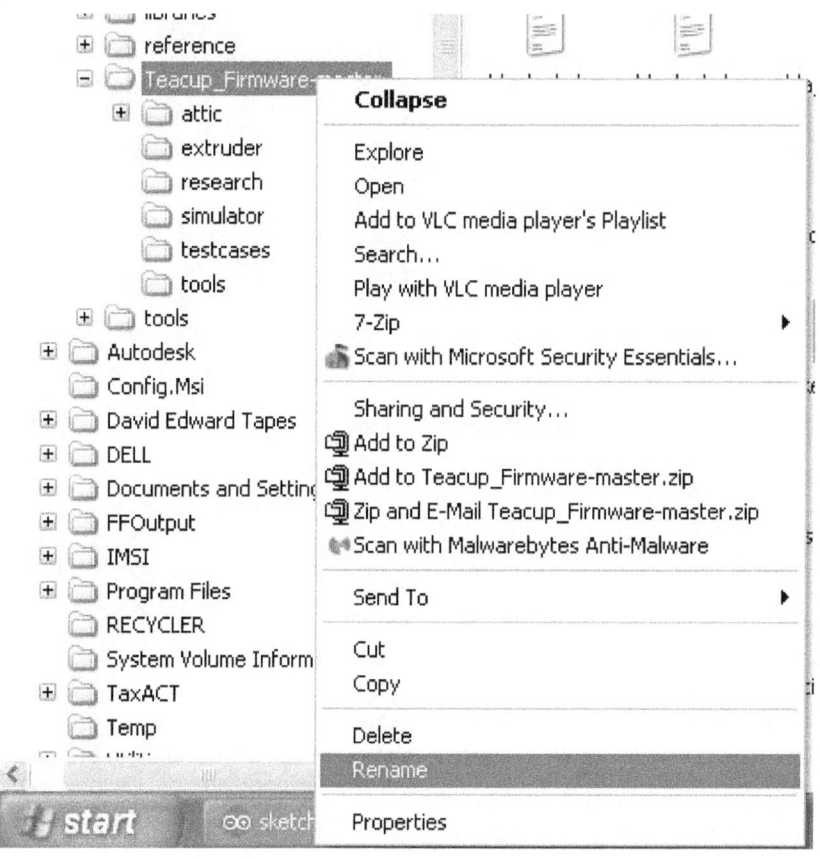

Rename "config.default.h" or "config.ramps-v1.3.h" to the "config.h". This decides what configuration (Ramps) you are going to be using. If you are going to be using mechanical end stop switches then you will need to edit config.h using notepad to remove the "//" that is in front of "#define USE_INTERNAL_PULLUPS". Otherwise you will need to add pull up resistors for the end stop switches to work properly.

Rename "ThermistorTable.double.h" to "ThermistorTable.h". This sets up how many thermistors you will be using in your 3D printer.

Load the Arduino interface software. Configure the Arduino interface for your Arduino Mega (Probably Arduino Mega 2560) and your serial port. Then select "open" and load the "Teacup_Firmware.pde". The teacup firmware will be found in that directory that you unzipped and renamed earlier.

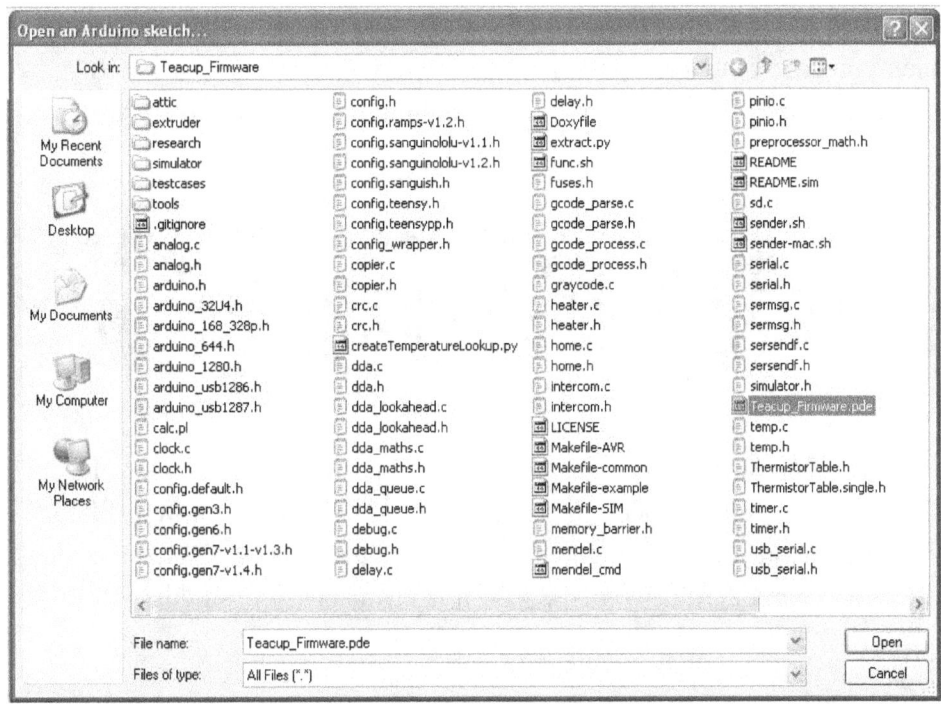

Once the firmware is loaded you should see several tabs like what is visible in the next picture. In fact these tabs run completely off the right side. If you click on the arrow on the right side of the tabs, an expanded list will run all the way to the bottom of the screen!

You can now upload the Teacup_firmware to the Arduino Mega like what is visible in the next picture.

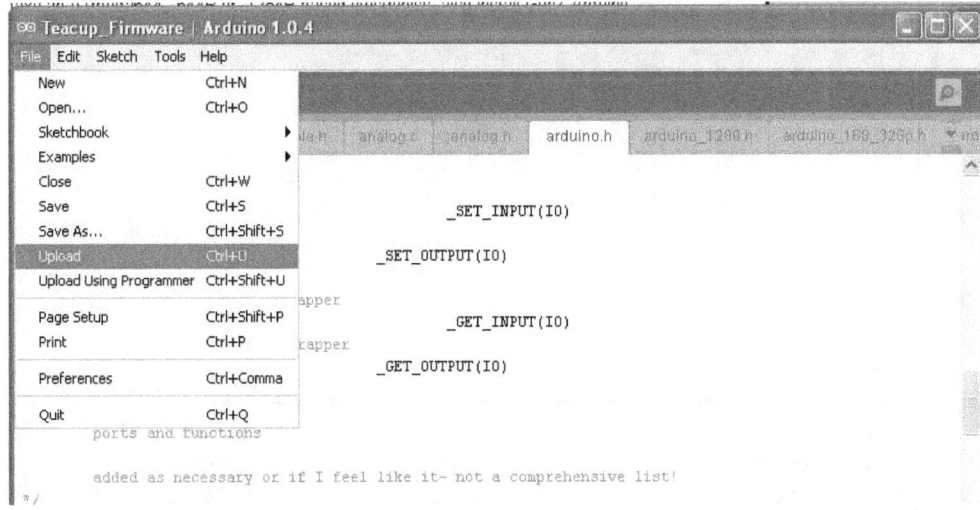

If you do not get any error messages at the bottom of the Arduino screen then you have now loaded the firmware into your Arduino Mega. A successful upload should look like something like this next picture.

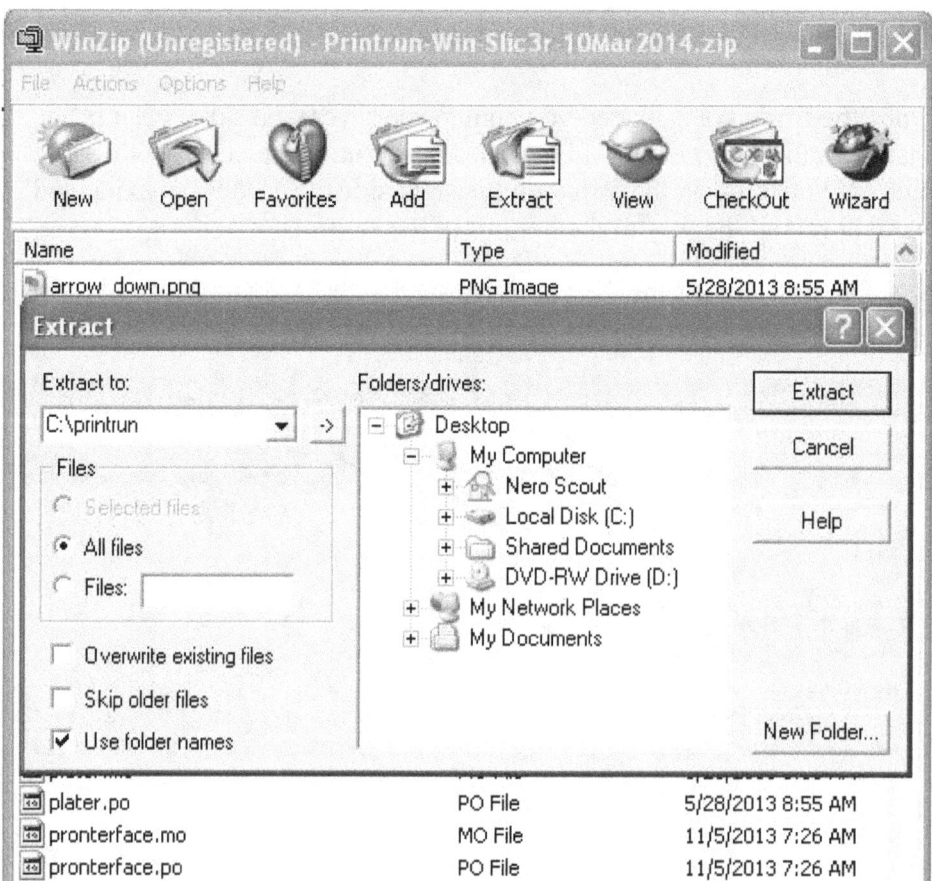

Next you will need the program for your PC to control the printer. I used Pronterface. You will need to find and download "PrintRun-Win-Slic3r". Then PrintRun is unzipped and extracted into a directory. I used the root directory of my hard drive or "C:" so it is always easy to find it.

Then you will need to locate that directory where you just extracted the files. Look for something called "Pronterface". Its icon is bright red with a ":-p" inside of it. Pronterface is the program for your PC to be able to run the 3D printer via the USB cable to the Arduino Mega.

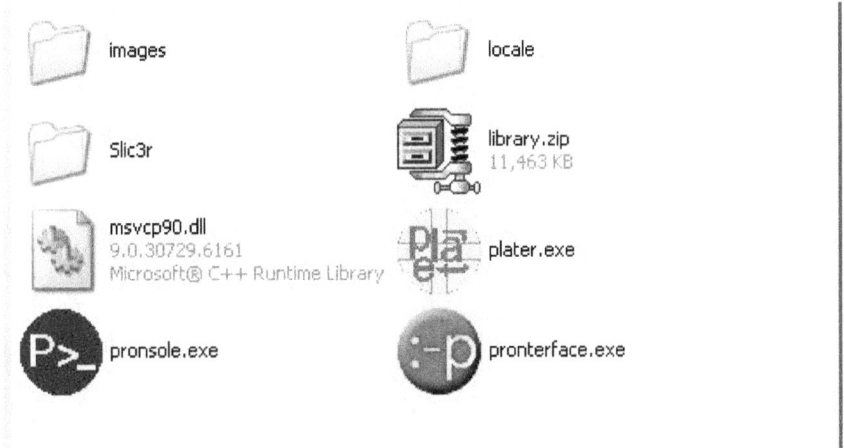

Pronterface looks like this once it is loaded and "Connected" to the machine.

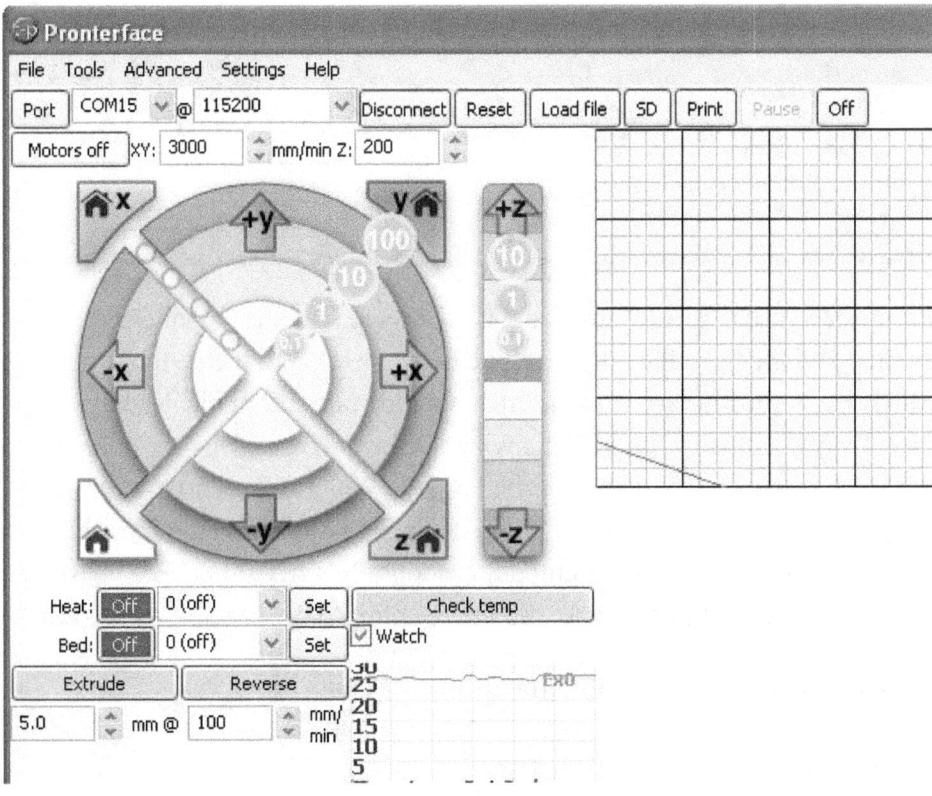

The big "Wheel" on the left side is the same thing as a CNC Pendant. You can choose the step size and the axis to move. You can also home each of the axes by selecting the icon with a picture of a "home" in it. Do not do that yet! You can now start to test out and calibrate your 3D printer.

Note that in Pronterface Under "settings" and "options" you can also adjust your communications settings, your temperature settings and your manual feed rates.

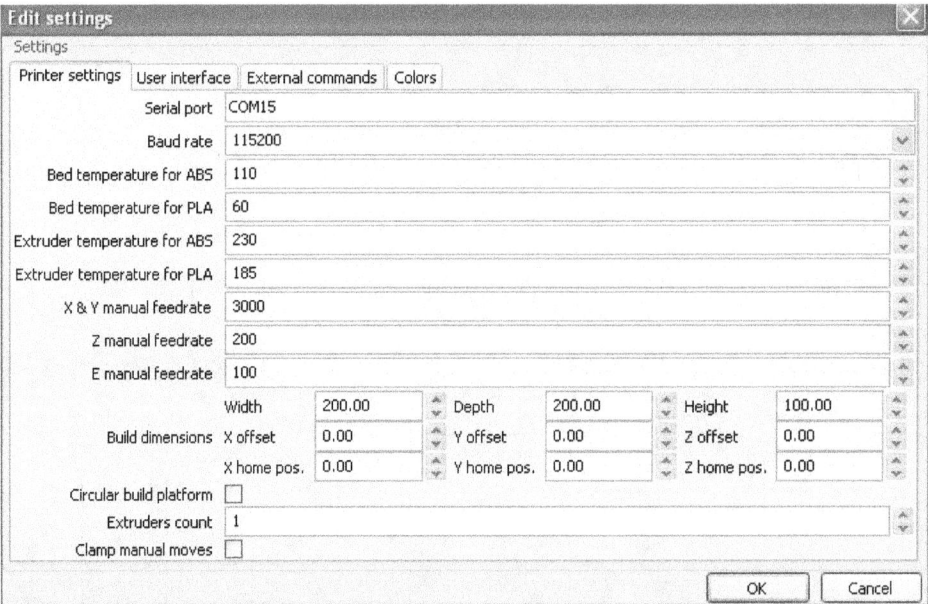

Pronterface uses "Slic3r" to slice the 3D image into slices that can be created and then produce the machine codes needed to create those slices. Slic3r has a configuration wizard that should start automatically but you can always manually run it. This wizard allows you to set your firmware type to "Teacup". Then set your bed Size to 200 mm by 200 mm. Next set your nozzle size to .4 in my case. Then set your filament diameter to 1.75 mm. Set the extrusion temperature to 220 degrees for ABS. Last of all set your bed temperature to 100 degrees for ABS.

This is what the Slic3r Configuration wizard looks like.

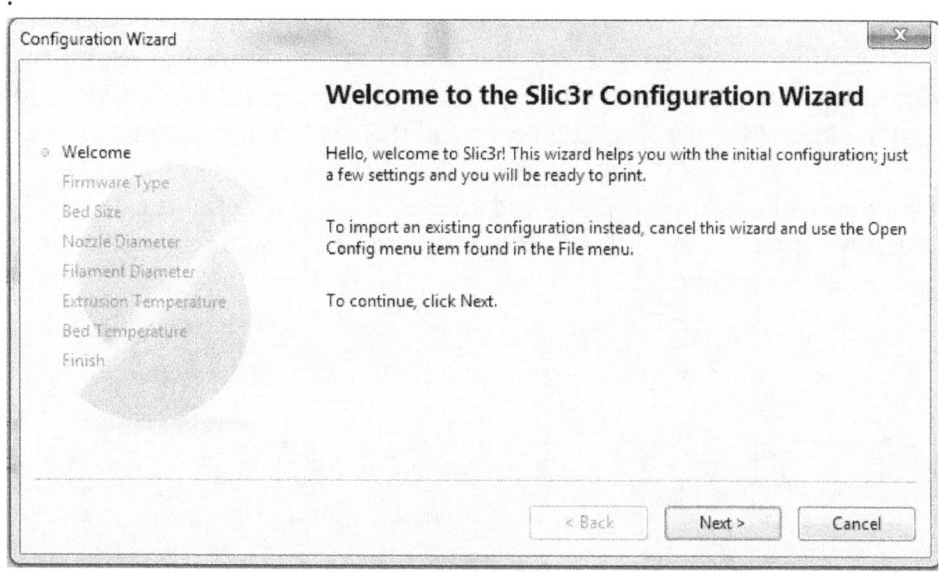

You can then verify your settings by looking at your filament settings and printer settings tabs as seen in the picture below.

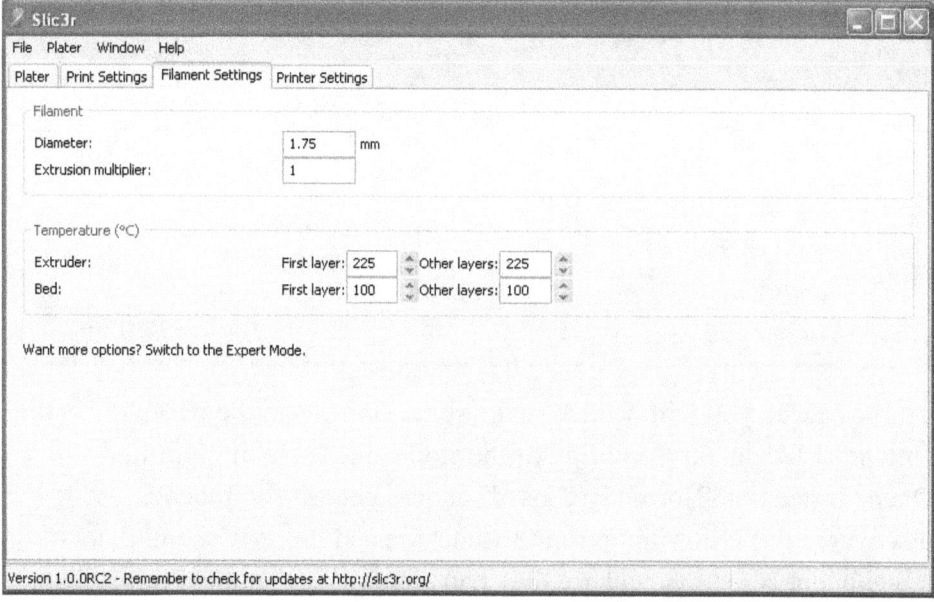

With all the software configured you can now start to set up and test out your 3D Printer.

Upgrading to Marlin Firmware

Marlin offers several advantages over teacup as your firmware running in the 3D printer. Marlin supports a LCD display, a pendant like mechanical jogging knob, SD memory, and best of all it is about ten times faster!

First you will need to download and extract Marlin. The file is called "Marlin-Marlin_v1.zip". I extracted it into my Arduino directory as you can see in this picture.

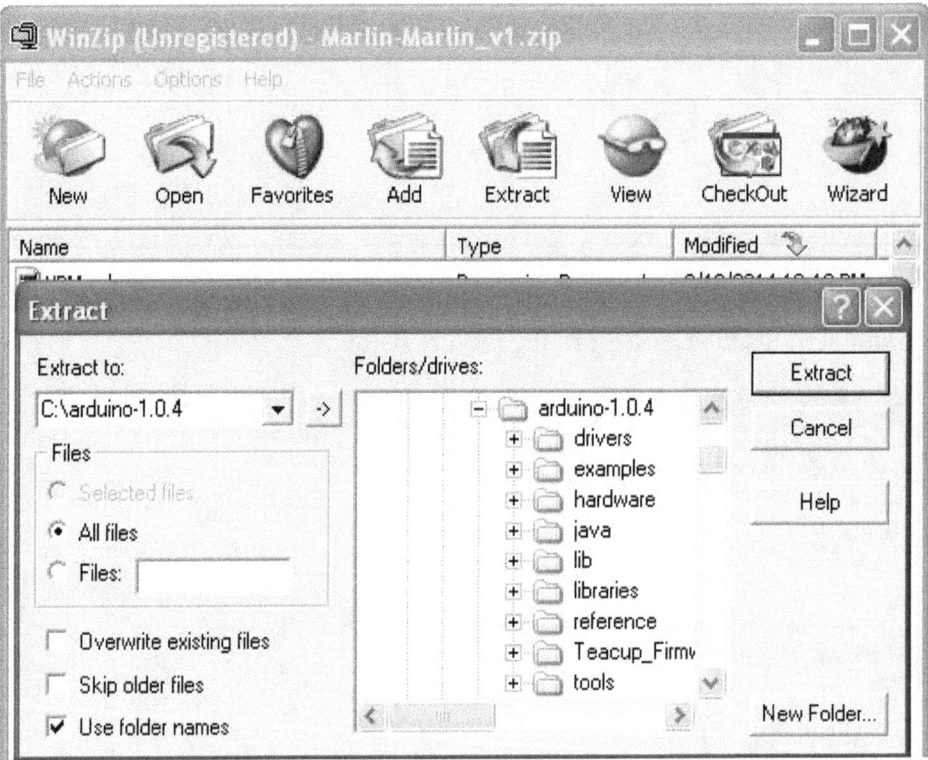

To run Marlin you will need to make some changes in the wiring or in the software. I changed the wiring of the stop switches to be "Normally Open" instead of "Normally Closed" as was needed for Teacup. I discovered that by trying to run the machine and the axis's would not move in one direction because they had already reached the limit switch. There were also a number of software changes that were needed as the machine kept crashing against the stops.

Once Marlin is extracted load up the Arduino IDE and from there load the

Marlin.PDE firmware.

Like Teacup, Marlin has many tabs. The configuration editing is done by editing "configuration.h". This file is a lot easier to find than config.h was in Teacup, it is just three tabs across.

The first thing you need to do is to define the motherboard as type "33" that is RAMPS 1.3 / 1.4. It should say "#define MOTHERBOARD 33".

Then there are the home direction settings. We will use the Mendel settings for the X, Y and Z axis. For the extruder we will use the "direct drive" settings. These settings are the opposite of the default settings.

```
Marlin   BlinkM.cpp   BlinkM.h   Configuration.h §   ConfigurationStore.cpp   ConfigurationSto

// For Inverting Stepper Enable Pins (Active Low) use 0, Non Inverting (Active High)
#define X_ENABLE_ON 0
#define Y_ENABLE_ON 0
#define Z_ENABLE_ON 0
#define E_ENABLE_ON 0 // For all extruders

// Disables axis when it's not being used.
#define DISABLE_X false
#define DISABLE_Y false
#define DISABLE_Z false
#define DISABLE_E false // For all extruders
#define DISABLE_INACTIVE_EXTRUDER true //disable only inactive extruders and keep ac

#define INVERT_X_DIR false    // for Mendel set to false, for Orca set to true
#define INVERT_Y_DIR true     // for Mendel set to true, for Orca set to false
#define INVERT_Z_DIR false    // for Mendel set to false, for Orca set to true
#define INVERT_E0_DIR true    // for direct drive extruder v9 set to true, for geared
#define INVERT_E1_DIR true    // for direct drive extruder v9 set to true, for geare
#define INVERT_E2_DIR true    // for direct drive extruder v9 set to true, for geared

// ENDSTOP SETTINGS:
// Sets direction of endstops when homing; 1=MAX, -1=MIN
#define X_HOME_DIR -1
#define Y_HOME_DIR -1
#define Z_HOME_DIR -1
```

There are some changes that need to be made in the "movement settings". To figure out the X and Y axis settings you can take the default 78.74 and multiply that by 20/16 to get 98.425. That is because I have 16 teeth gears instead of 20 teeth gears on those two axes. However the setting should have been 100 as it was for Teacup. The number 100 is the normal setting for 200 steps per revolution times 16 micro-steps divided by a belt pitch of 2 times 16 teeth.

The Z axis setting of 2267.716 is the standard setting for a 5/16 inch by 18 teeth per inch Z axis leadscrew. Although according to my tests 2267 was a more accurate number. The 5 mm leadscrew uses 4000 and the 8mm leadscrew uses 400 (it has two sets of threads)

The 110 setting for the extruder was determined by trial and error. I would try a number then load Pronterface and extrude or retract 10 mm and then measure the amount that the filament moved. The order of the steps per unit numbers is X, Y, Z, then E.

Note in the picture below that I also reduced the max acceleration numbers from 9000 to 5000. Also the Z axis homing feedrate is reduced to 4*60, it might have to be set even lower if the Z axis homing results in the steppers making loud noises and/or failing to move properly.

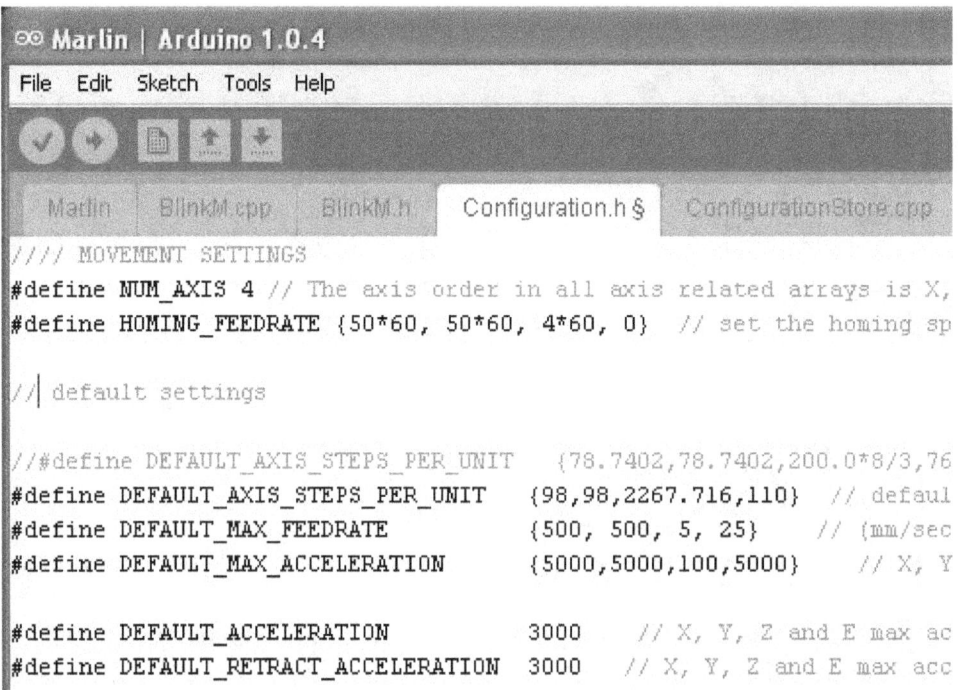

One of the things that you will eventually need to do is to print a 20 mm by 20 mm by 10 mm tall test block and verify that your X, Y and Z settings are correct.

Last of all there are the temperature settings. The important change there is to set the temperature bed to "one" thus enabling the heat bed.

```
Marlin     BlinkM.cpp     BlinkM.h     Configuration.h §     ConfigurationStore.cpp
//     1k ohm pullup tables - This is not normal, you would have to hav
//                          (but gives greater accuracy and more stabl
// 51 is 100k thermistor - EPCOS (1k pullup)
// 52 is 200k thermistor - ATC Semitec 204GT-2 (1k pullup)
// 55 is 100k thermistor - ATC Semitec 104GT-2 (Used in ParCan & J-Hea
//
// 1047 is Pt1000 with 4k7 pullup
// 1010 is Pt1000 with 1k pullup (non standard)
// 147 is Pt100 with 4k7 pullup
// 110 is Pt100 with 1k pullup (non standard)

#define TEMP_SENSOR_0 1
#define TEMP_SENSOR_1 1
#define TEMP_SENSOR_2 0
#define TEMP_SENSOR_BED 1
```

After these changes are made and saved you can then upload the firmware to the Arduino Mega that is located inside the 3D printer.

When I loaded Pronterface, but it would not connect to the printer until I changed the Baud rate to 250000. That setting can be found right after the port setting in the upper left area of Pronterface as can be seen in the next picture.

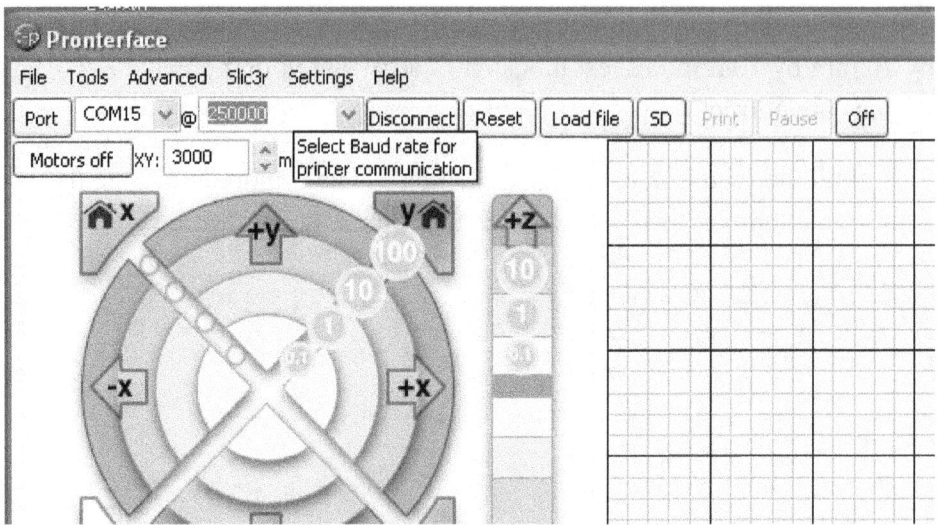

Even after all of these changes there were some more wiring changes to make. The Z axis was inverted, moving in the opposite direction so I reversed the two Z connections at the RAMPS board. When I tested the extruder it was also inverted so I reversed its connection to the RAMPS board as well.

When all of this was done I loaded a 10 mm linear bearing holder and let the machine make it. After the first layer was a mess I stopped the machine and slowed it down in slic3r. The machine was running faster than it could extrude. Then I tried again and in less than two hours I had a bearing holder that took about 10 hours for Teacup to make. In fact the machine runs so much faster that some screws came loose. Now I see why they recommend that you use lock washers in the screws!

Chapter 8
Calibration and Operation

Using Pronterface the end stop switches can be checked out by issuing a "M119" command in the bottom right corner as can be seen in the next picture. Check to see that they are showing "open" then you can manually press each switch one at a time and at the same time reissue the M119 command to see if it says "triggered".

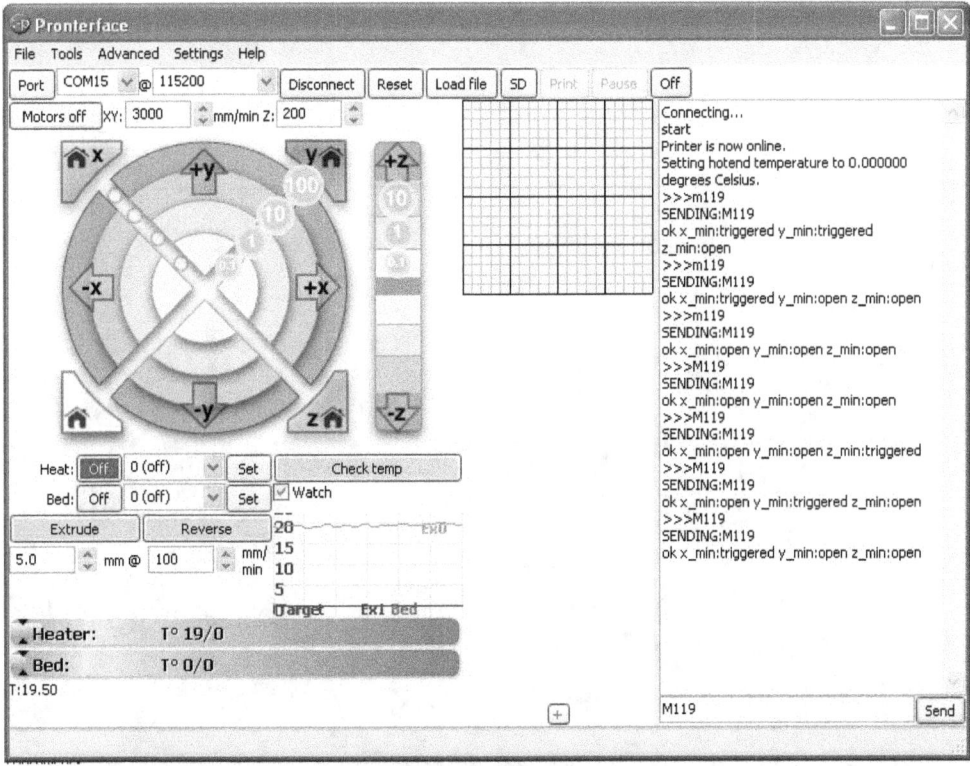

The axes appear to be reversed from what the operator sees when he looks at the machine. Instead, try to imagine that you are the print head and rotate yourself around to being located behind the machine looking down at the print surface. Now you will see that the positive X movement goes to the right and positive Y movement will move the Y axis towards you so you will print at the top of the surface. Basically a X,Y coordinate of 0,0 is located in the back left corner of the heat bed.

So from the viewpoint of standing in front of the machine the X axis negative is on the right, positive X is moving to the left. The Y axis should move away from you for a positive movement and towards you for a negative movement. If these are not correct then turn off power to the motors, then unplug the offending motor and reverse the plug. Then power it back up and it should now move correctly.

Here is a picture of the 3D printer with arrows to show how the axes should move if they are properly connected.

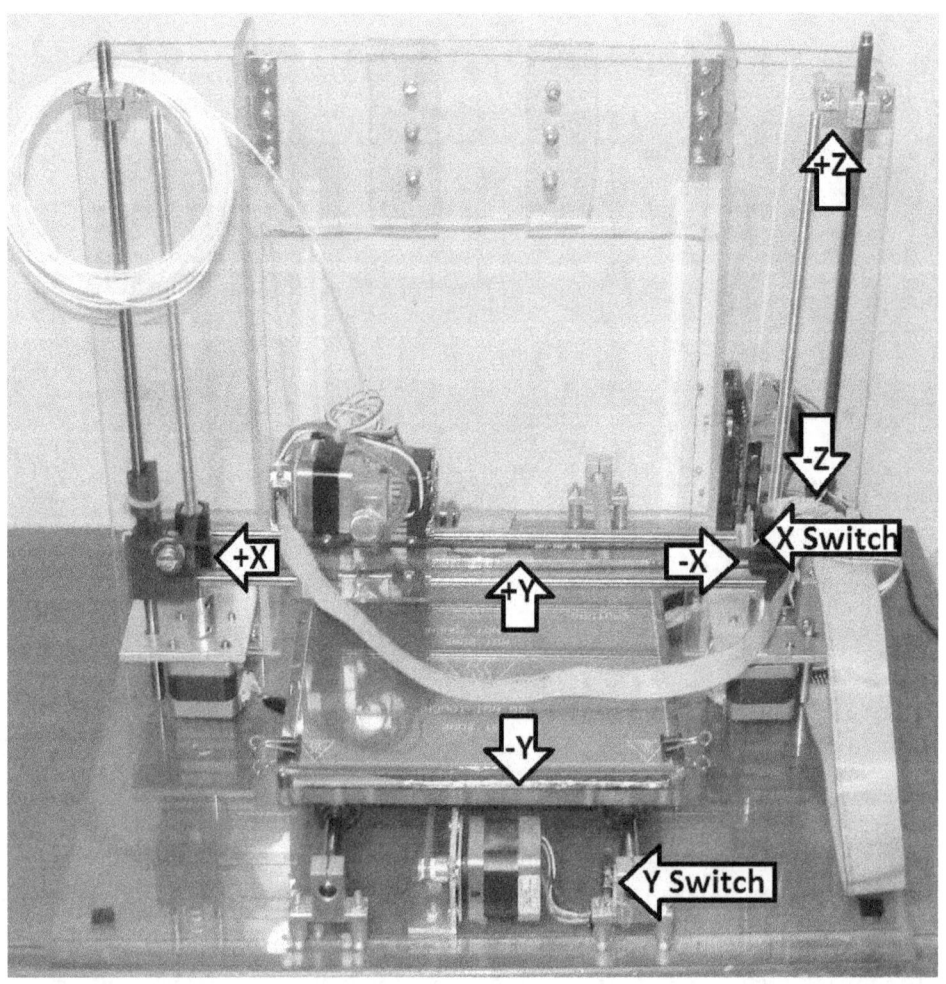

Since the 0,0 orientation changes between the machines lets look closer at it. The Mendel 90 has the limit switches at the right for the X axis and front for the Y axis. So the X axis moves left for positive X movement. From the front view this looks backwards as the 0,0 reference is 180 degrees out from the front view. It is in the back right corner.

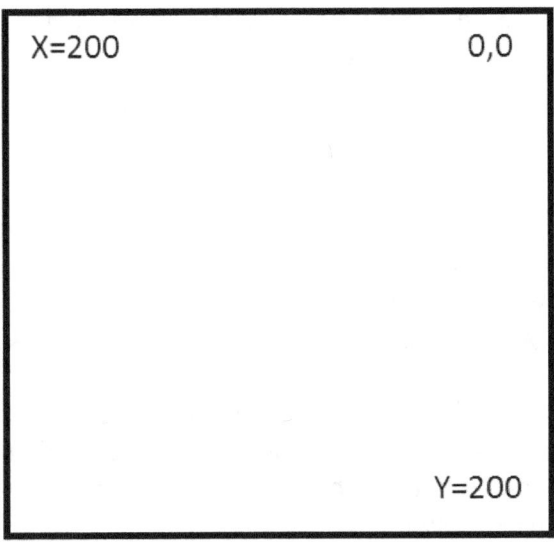

The Geeetech i3 has the limit switches at the left for the X axis and the back for the Y axis. So the X axis moves right for positive X, but the Y axis moves forward for positive Y movement.

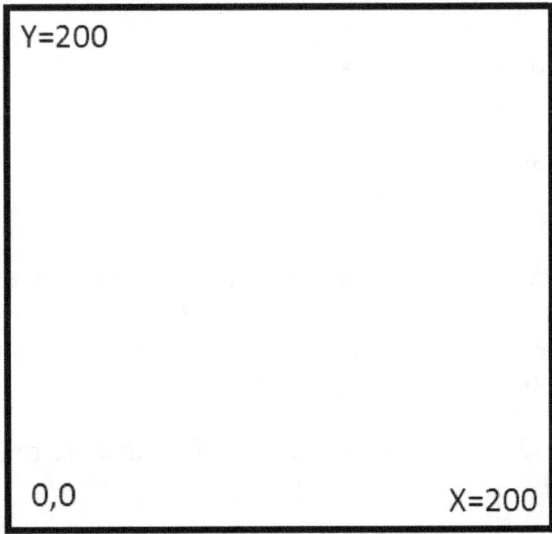

Now it is safe to try out each of the "Home" buttons and see if the machine goes to its home position. If something goes wrong you will need to turn off power to the stepper motors to prevent the machine from crashing. Plugging the 12 volt AC adapter into a power strip provides one way to quickly turn it off.

The next step is to calibrate the axes. For the X and Y axis center the axis first, then take out a ruler and select a 100 mm movement. Measure the movement and it should be a little less than four inches as 100 mm is 3.93 inches.

My results of this test were as follows on the first test:
X axis 100 mm was 3.25 inches.
Y axis 100 mm was 3.25 inches.
Z axis 100 mm was 5.75 inches.

After some research I discovered that the normal number of teeth on the gear for the X and Y axis is 20 teeth and I had purchased and installed 16 teeth gears. So I had to go to config.h in the Teacup_firmware files and change the following lines.

// Old Values 20 teeth and 8mm for Z

//#define	STEPS_PER_M_X	80368
//#define	STEPS_PER_M_Y	80368
//#define	STEPS_PER_M_Z	3333592

// New Values 16 teeth and 5/16 for Z

#defineSTEPS_PER_M_X	100000
#defineSTEPS_PER_M_Y	100000
#defineSTEPS_PER_M_Z	2267716

When you load the teacup firmware in the Arduino IDE the config.h file is hard to find. There is a row of tabs to the right of "Teacup_Firmware" and to the right of the tabs there is an arrow. Click on that arrow and a list drops down that will have config.h in it.

The next picture shows the extended list of files that are found in the teacup firmware.

Then the changed files are uploaded to the Arduino Mega and the calibration tests were run once again. This time the measurements were correct. The next things to calibrate are the Z axis and then the extruder.

To calibrate the Z axis, first home all three axes. A piece of paper should be able to slide between the print head and the surface of the glass on the heatbed. Now turn off power to the stepper motors. Manually slide the X axis all the way to the left. See if a piece of paper still just so fits under the print head (Extruder hot end). If not, manually rotate the left stepper so it is the same as the right side was. Next move the Y axis to its other end and test with the paper again. Then move the X back to its home and test one last time. My right front corner needed a thin washer to make the surface level for some reason. Your Z axis is now properly calibrated.

To calibrate the extruder first mark some filament at 10 mm intervals. Use the "extrude" button to feed the filament in and "reverse" to feed the filament back out. You might have to remove the hot end to do this. See if 10 mm of filament moves in when it is requested. Mine only moved a couple of millimeters. It is time to go figure everything out and adjust the settings in config.h once again!

```
// Old Value
//#define STEPS_PER_M_E              11036
// New Value
#defineSTEPS_PER_M_E                110360
```

Here is what a marked filament looks like, it does not have to be exact.

Once your filament is feeding properly, setting up the heat bed is next. My heat bed did not even register in Pronterface. In fact, it came up as if the temperature was stuck at zero degrees. I checked everything out with the voltmeter and it was connected properly. When I plugged the thermistor into the extruder jack it registered the temperature. Then I went back into config.h and discovered not one, but that there are a total of three lines that need to be changed.

Remove the slashes in front of the heat bed temp sensor in each of these three locations.

```
//            name      type      pin      additional
```

DEFINE_TEMP_SENSOR(extruder, TT_THERMISTOR, AIO13, THERMISTOR_EXTRUDER)

//DEFINE_TEMP_SENSOR(bed,   TT_THERMISTOR, AIO14, THERMISTOR_EXTRUDER)

// NOTE: these pins are for RAMPS V1.1 and newer. V1.0 is different

```
//        name    port   pwm
```

DEFINE_HEATER(extruder, PB4, 1)

//DEFINE_HEATER(bed,   PH5, 1)

#defineHEATER_EXTRUDER HEATER_extruder

//#define HEATER_BED HEATER_bed

Then resend teacup to the Arduino Mega, and then load Pronterface and see if the heat bed is now working. Once all these things are working you can turn on the heat and try to make something. There are articles available on what the first layer should look like. It has to stick to the glass or tape and be flattened slightly.

As you can see in the next picture my first extrusions did not turn out very well. The first layer was OK but after that the extrusion got really thin and even skipped. When I put my hand on the filament it was only taking in really small amounts. Then if I added some downward pressure sometimes as much as 1/2 inch of filament fed into the extruder.

So I powered it all down and tested out the wiring to the extruder stepper motor. You should meter a short between pin one and pin two as well as a short between pin three and pin four. The motor wiring checked out fine. Then I looked at the trimmer resistors on the RAMPS board. The problem was there in plain sight. All of the trimmers were set to the middle of their range; the flat spot was facing down, except for one of them – the extruder. They should all have the flat side down except for the Z axis, it should be rotated clockwise about 90 degrees as it runs two motors.

This is a picture of what the RAMPS trimmer resistor should look like.

While on the subject of the stepstick boards there are three jumpers hidden underneath them. They should all be "on" for normal 1/16 steps. If you upgrade the stepstick boards then you might have to change the jumper settings. Here is a chart showing what the jumpers settings do.

Step size jumper settings for A4988 stepstick

ms1	ms2	ms3	step size
off	off	off	full step
on	off	off	half step
off	on	off	1/4step
on	on	off	1/8step
on	on	on	1/16step (Normal setting)

Step size jumper settings for Drv8825 stepstick

ms1	ms2	ms3	step size
off	off	off	full step
on	off	off	half step
off	on	off	1/4 step
on	on	off	1/4 step
off	off	on	1/16 step
on	off	on	1/32 step
off	on	on	1/32step
on	on	on	1/32step

With those changes I finally had my first successful extrusion. It looks a little rough because my extruder keeps extruding even when it is told to stop. The excess plastic can be removed with a razor knife and a file. The left picture is ABS and the right picture is PLA. Notice that the PLA is much easier to work with and it produces a better looking product.

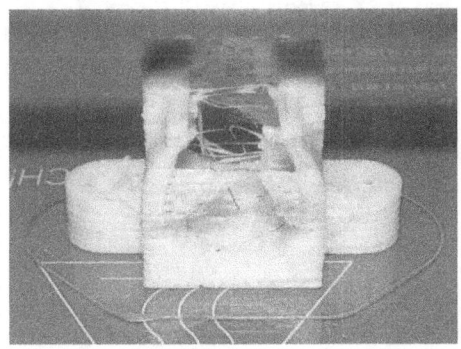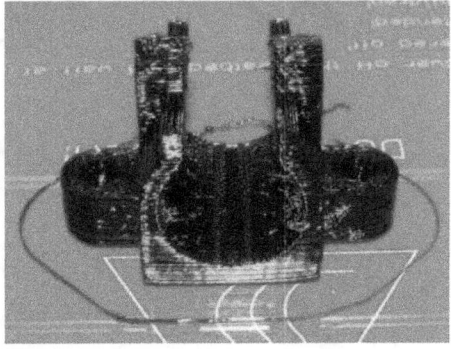

Some of the problems with printing include the need to clean off the glass. If the glass is not really cleaned the object being printed will come loose. Some recommend using acetone to treat the clean glass. Also hair spray will help to keep things attached to the glass. If it does come loose you might get something that looks like a disaster area as it will keep on trying to print on what is not there. Here is a picture of one such "creation" that I made.

Another issue in printing a lot of things is setting the Z axis. The mechanical switch tends to not be exact enough and an optical switch would be better. I have found that it is best to calibrate the machine so that the z axis "home" just so touches the glass but does not cause the motors to skip.

Before you start printing something you can use Pronterface's "edit" command to look at the beginning of your file to see if the settings are completely correct for your printer. Look for things like the correct filament size and the correct nozzle size.

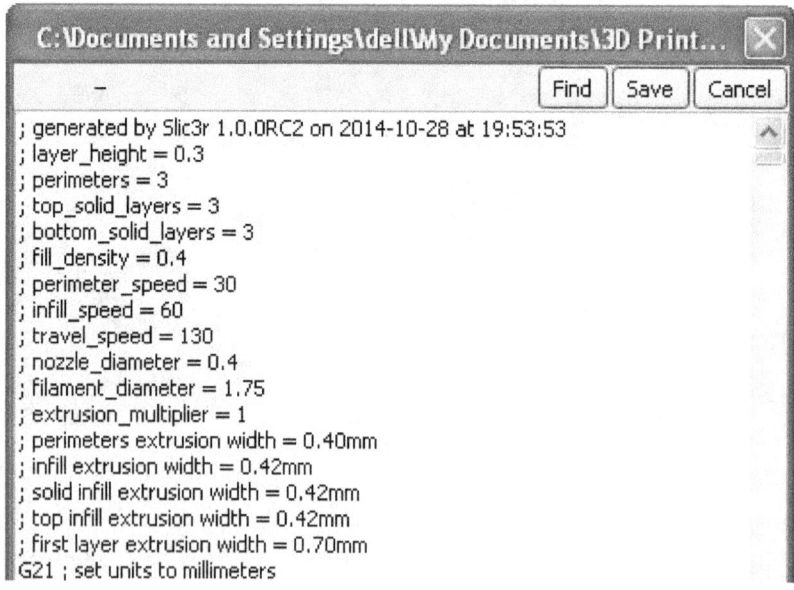

Chapter 9
Filament Spool Holder Options

After using some small rolls of filament, I ordered a bigger spool of filament. The problem with the big spool is that there needs to be a way to mount the spool so that it can rotate freely. Some people mount the spool of filament behind the gantry, but that can lead to feeding problems. My solution was to mount the spool on the left side of the front of the 3D printer.

What I did was to mount a three inch by two inch piece of Plexiglas that has a 1/2 inch hole in the center on the back side of the gantry. I used the screws that hold the Z axis 8 mm rod holders in place to hold the new piece of Plexiglas in place.

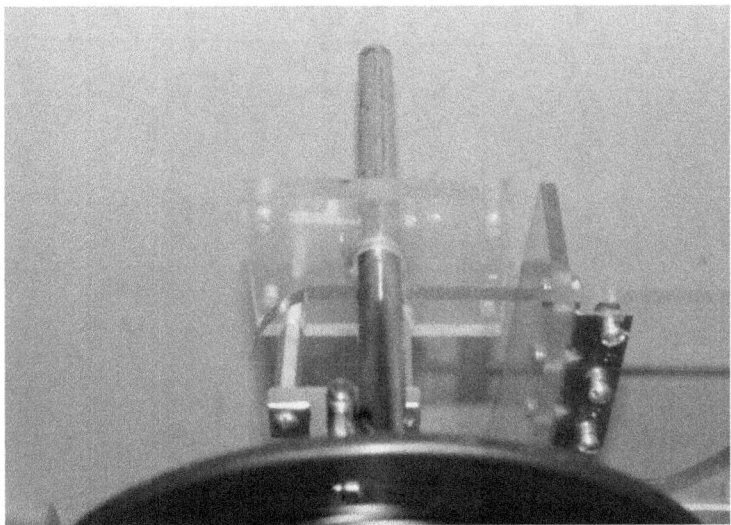

Then for the front of the spool holder I used an 18 inch by three inch piece of Plexiglas that already had a couple of 1/2 inch holes in it. At one time this had been the front of my home made CNC machine that is discussed in another book.

Through the center of the spool there is a 1/2 inch by 15 inch rod that was from a HP printer. Two pipe straps, one at each end, hold the rod in place. Here is a picture of my filament spool holder.

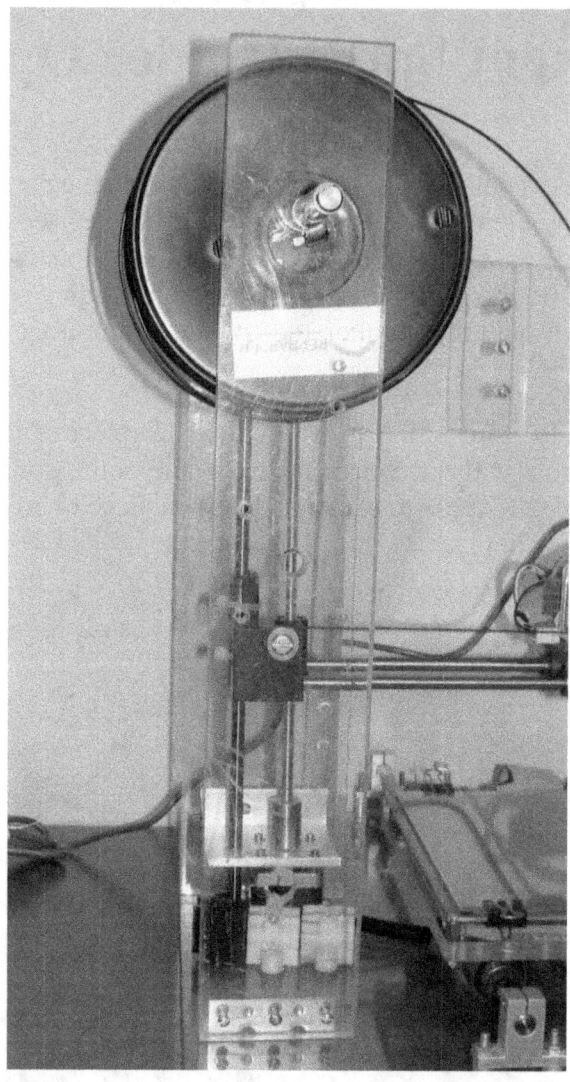

Another way to add a spool holder is to extend the gantry braces. To do that you will need two pieces of Plexiglas about nine inches by three inches in size. You will need to drill a 1/2 inch hole about one and a half to two inches from the top and centered front to back. Then you will need to drill three holes 1/2 inch, 1.5 inches and 2.5 inches from the bottom and 1/2 inches from the front edge. These holes then line up with the gantry brace screws. Here is a drawing of one of the brace extensions.

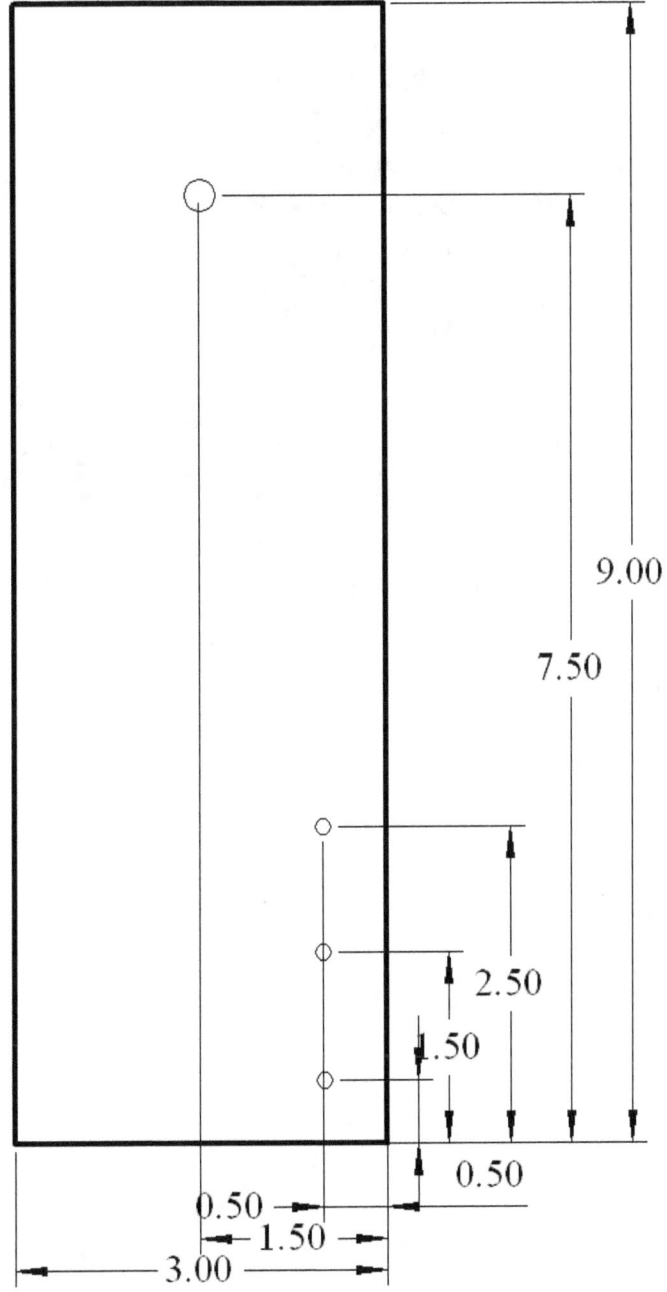

Once again a 1/2 inch by 12 to 16 inch rod goes through the ends and through the center of the spool. Then the rod is held in place with two pipe straps one at each end.

Here is a picture of what this reel or spool holder setup looks like.

This spool holder design makes the machine nearly two feet tall. However it does a better job of getting the spool of filament out of the way.

These gantry extensions can also be used for holding the LCD display that will be introduced in the next chapter.

Chapter 10
Adding LCD Display, Encoder, and SD Memory

You can purchase a LCD board adapter for a RAMPS powered 3D printer on eBay. If you want to make your own it is a bit tricky because the order of the pins does not match up. It took me three tries at soldering together the adapters to get it done right. I used a two pin connector for power on pins one and two. Then I used an 8 pin connector to connect to pins 11 to 18. The LCD only uses pins 13 to 18. Some of the other pins are used for a "jogger" or "encoder" type of device and for a buzzer.

This is a listing in the order of the LCD pins.

LCD	Aux 4	Use
1	2	Gnd
2	1	5V
3	Trimmer	Contrast
4	18	RS -> D16
5	Gnd	R/W -> Gnd
6	17	Enable -> D17
7	NC	Data0
8	NC	Data1
9	NC	Data2
10	NC	Data3
11	16	Data4 -> D23
12	15	Data5 -> D25
13	14	Data6 -> D27
14	13	Data7 -> D29
15	NC	Backlight Anode -> 5V
16	NC	Backlight Cathode -> Gnd

This next listing is in the order of the Aux 4 Jack located on the RAMPS.

Aux 4	LCD	Use
1	2, 15	5V
2	1, 5, 16	Gnd
3		
4		
5		
6		
7		
8		
9		D37 – Optional Buzzer
10		D35 - Encoder Click
11		D33 - Encoder Direction
12		D31 - Encoder Direction
13	14	D29 -> Data 7
14	13	D27 -> Data 6
15	12	D25 -> Data 5
16	11	D23 -> Data 4
17	6	D17 -> Enable
18	4	D16 -> RS

You might want to add a 10 pin connector in between the Aux4 18 pin connector and the 16 pin LCD connector. If you buy a kit it most likely has this connector already wired in. Here is the pinout of the optional 10 pin connector.

Here is a picture of the 10 pin ribbon cable being adapted to the 8 pin and 2 pin connectors to plug into the 18 pin plug on the RAMPS board.

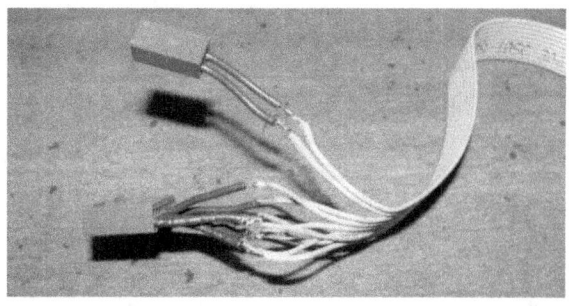

Here is the schematic diagram for wiring up a typical 16x2 LCD. The same design works for almost all text based LCD's.

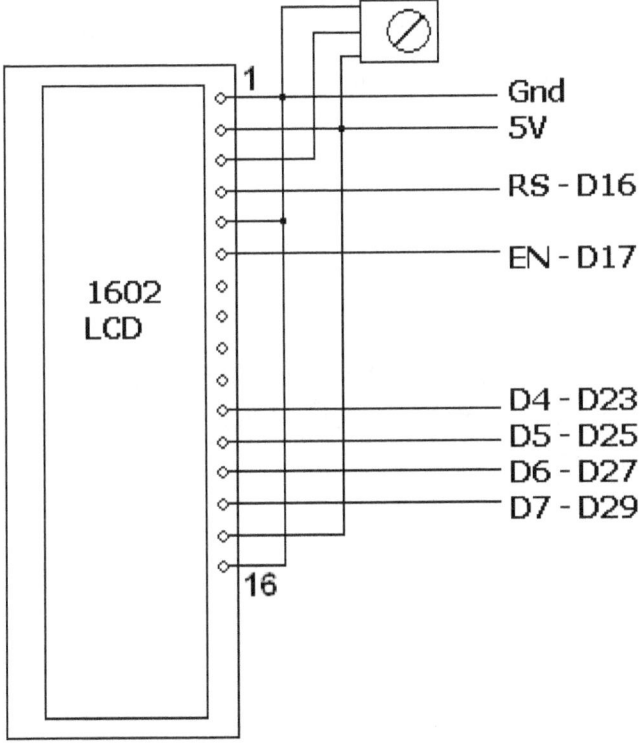

To reduce the amount of wires running to the LCD I consolidated the power and ground wires with some 30 gauge wire jumpers. Then I replaced the trimmer for contrast with a jumper to ground as can be seen in the next picture.

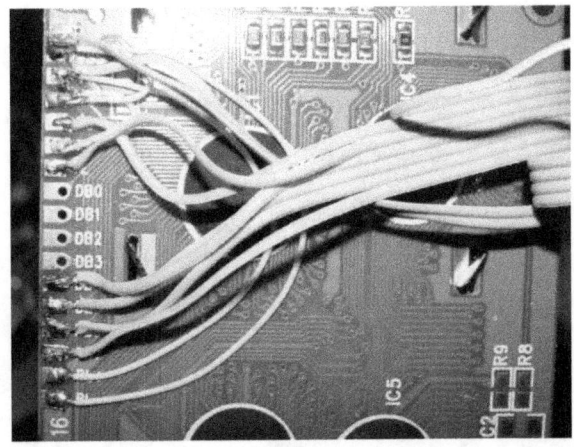

Once you have the LCD wired up and powered up you should see a row or two of black boxes. To enable the LCD in software you will need to make one or two changes. The first change is to enable the LCD support. In Configuration.h after the line "//LCD and SD support" remove the slashes in front of "//#define ULTRA_LCD".

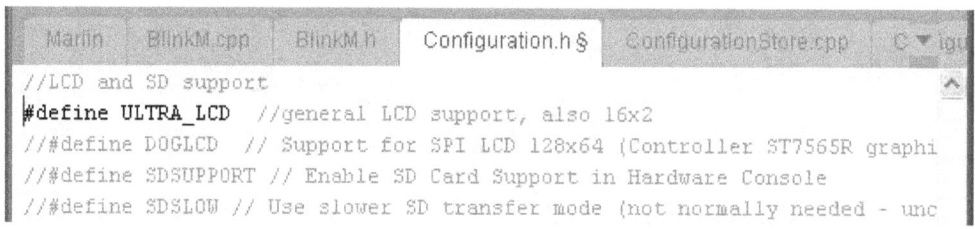

```
//LCD and SD support
#define ULTRA_LCD   //general LCD support, also 16x2
//#define DOGLCD   // Support for SPI LCD 128x64 (Controller ST7565R graphi
//#define SDSUPPORT // Enable SD Card Support in Hardware Console
//#define SDSLOW // Use slower SD transfer mode (not normally needed - unc
```

To set up the size of the LCD, if it is other than 16x2, you will need to locate the "Ultra_LCD" line definitions. Look for the following lines and make your changes.

#ifdef ULTRA_LCD
   ……
#else
  #define LCD_WIDTH 16
  #define LCD_HEIGHT 2
 #endif

Change the LCD_WIDTH and LCD_HEIGHT numbers to match your LCD size.

Note that to use a 40x2 LCD you need to configure it as a 20x4 LCD. When using a 40x2 LCD the right side of the LCD display shows the bottom two lines.

Here is a picture of a typical 16x2 line LCD display.

This next picture is of a typical 40x2 line LCD configured as 20x4.

Here is a picture of a 2004 with a 20x4 line LCD display.

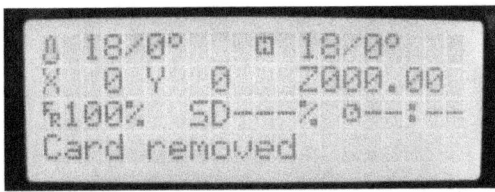

Next we will add a rotary encoder. The easiest way to do that is to enable the "REPRAP DISCOUNT SMART CONTROLLER". This will also enable the SD card adapter.

The encoder connects with a wire that connects the center of the set of three pins and then one of the set of two pins then goes to ground on the LCD display. Then there are three wires to connect that come from the RAMPS. The encoder direction wires go to either side of the set of three pins. The click wire goes to the other one of the two pins. This can be seen in the next picture.

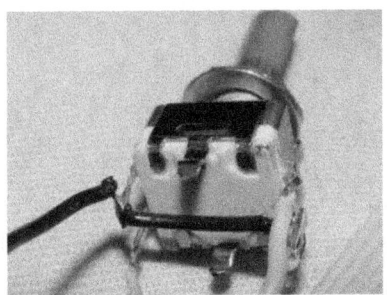

Once it is wired up you can push in on the knob and the LCD display should change. Then you can rotate the knob to move up and down through the items on the LCD display. If the direction is backwards you need to reverse the encoder directions wires. Here is what the LCD display should now look like. Note that the SD card is working; it is just not being detected properly.

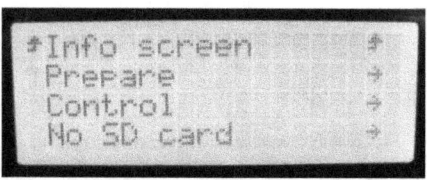

The parts for these updates are all available on eBay. Here is a typical 2004 or 20x4 LCD display.

Here is the rotary encoder switch.

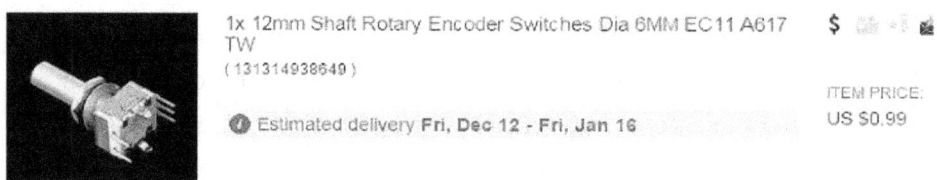

Here is the SD card adapter. It plugs into the 8 pin connector in the lower right corner of the RAMPS board with the memory slot facing out.

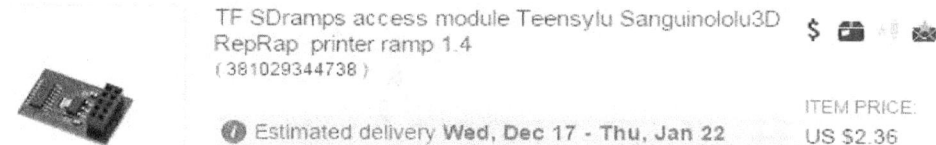

If you want to add an optional LCD display and do not want to wire up your own LCD, here is one that is found on eBay. The ribbon cables might be a little too short. I had to make longer ones for it.

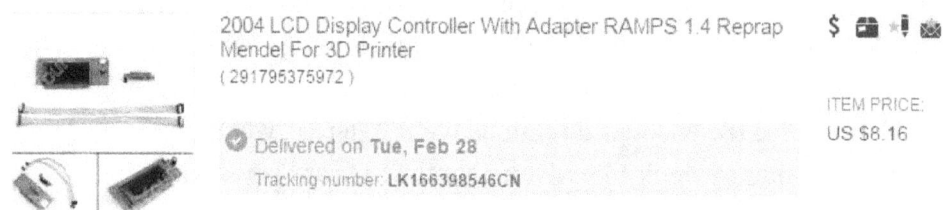

Chapter 11
Extruder Options

The extruder is the "heart" of a 3D printer. It is the one device that totally separates it from all other machines. If you buy a cheap extruder you will likely be frustrated with the results. Extruder problems make it difficult to get your 3D printer to work properly. This is an examination of the parts of the extruder, how to tell if they are built correctly and how to fix yours if it is having issues.

My 3D printer is very similar to the Mendel 90 design. I made my own motor mounts out of aluminum and for the guide rails supports I used standard aluminum rail holders. Like on the Mendel 90 I used a standard 15 pin plug to connect to the extruder. This plug makes it easier to change out the extruder. That is the 15 pin plug that is visible in some of the extruder pictures.

Here is a picture of my starter extruder. I thought it was a great deal when I bought it because it was an all metal design. I really did not know what to look for in an extruder at that time. It was designed to fit a "MakerBot" with a very low Z axis profile. We will call this "extruder one".

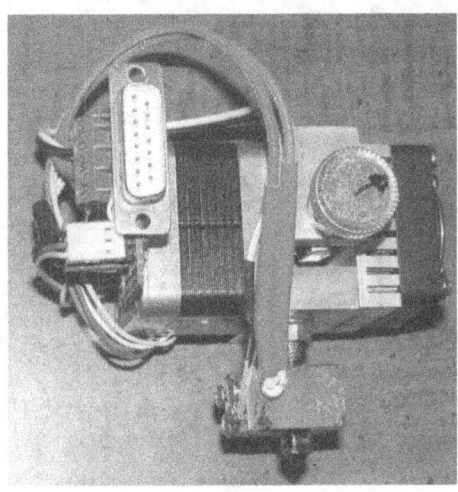

My first few creations had some voids or gaps in them. Sometimes the extruder would quit working altogether. This problem resulted in making what I call "Bricks" or worthless plastic pieces. The 10 mm linear bearing supports that I made on my 3D printer broke easily. Why? Because of inconsistent filament feed rates. Here is a picture of some of my "Bricks". I have made a lot more of them since then!

Take a close look inside of my extruder in the next picture. I removed the fan and the heat sink to take a look inside of the extruder. There are "Filament chips" lying at the bottom. The Teflon tube has risen up out of the hot end. The gear teeth are flat and flat teeth do not grip a round filament very well. The tension adjustment screw is not even spring loaded. I bought this extruder on eBay for $55. It is why they say "You get what you pay for".

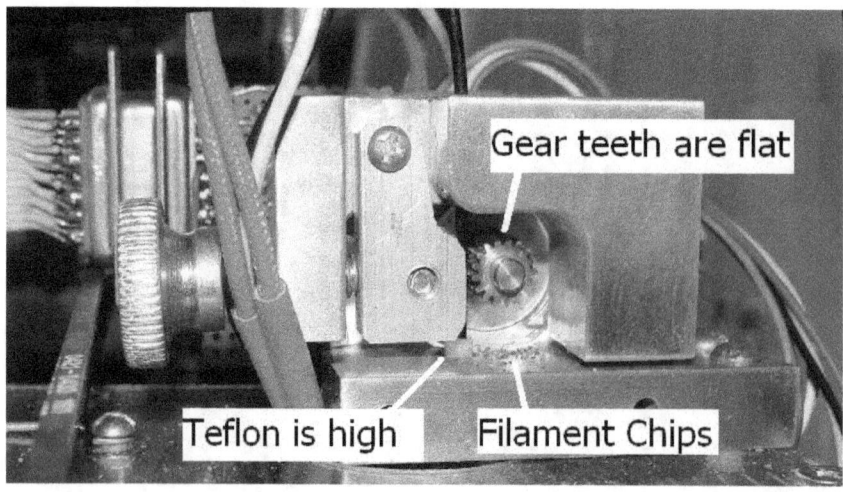

These extruder inadequacies resulted in the voids that I experienced in my parts. To resolve these issues I studied extruders and learned a lot of things about how they are made and how they should work.

For instance extruders for three millimeter filament are usually "geared" to give them a mechanical advantage resulting in a lot more pulling or pushing power. The bolt that is used to contact and move the filament is "hobbed" so that it has teeth that are bowed in to produce a curved gear to better grab the round filament. The tension bearing is held in place by a spring or by multiple springs to automatically compensate for varying filament diameters.

The smaller 1.75 mm filament can be used with the smaller extruders. These smaller extruders are usually called "direct drive" extruders. That is because the stepper motor gear directly drives the filament. The direct drive extruders have evolved to use a drive gear that is also concave, or "hobbed" and they are now spring loaded as well. There are two arrangements for direct drive extruders. One arrangement has the heat sink and fan located above the platform, like extruder one, and the other has the heat sink and fan located below the platform. Here is a picture of what a "hobbed" drive gear looks like.

The better extruder hot ends have several design improvements in them as well. On some models the metal threaded rod screws into some PTFE (A high temperature plastic insulator) that then has a mounting groove in it at the top. This groove fits into a matching groove in the cold end of the extruder or it can fit into a hole in the cold end and then it is held in place with two screws. A cooling fan can be added to help keep the heat from rising up into the extruder. If the threaded rod is not converted to PTFE, it can have aluminum washers or rings added as a heat sink to help it dissipate the heat. This heat sink is added to keep the heat out of the extruder's cold end.

If the Teflon insulator inside of the hot end does not fit tightly it can move up allowing a "Plug" to form below it. This plug happens when there is pressure down on the filament. Some of the molten filament moves up into any gap between the Teflon insulator and the nozzle. Then every time the filament is retracted this plug moves up pushing up on the Teflon insulator. Then either the Teflon moves up as well, or if it cannot move up the extruder jams. If the Teflon insulator cannot move up it might get deformed to the point that it can prevent the filament from moving up or down through it. To test this theory I heated the hot end to just above the melting point and quickly pulled the plug out. Here is a picture of the plug that came out of the hot end. Note that the plug is big enough that it extends up above the filament melting area, so the top of the plug would not be melted.

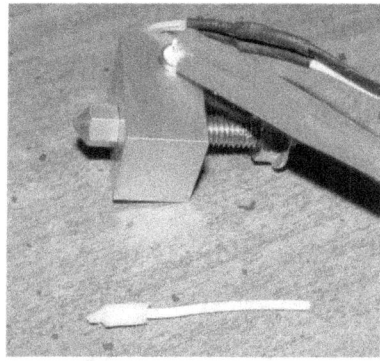

Here is a diagram of a typical direct drive extruder showing the cold end (at the top) and the hot end (at the bottom). The terms "PTFE" or "PEEK" are names for high temperature plastic with a 327 degree centigrade melting point.

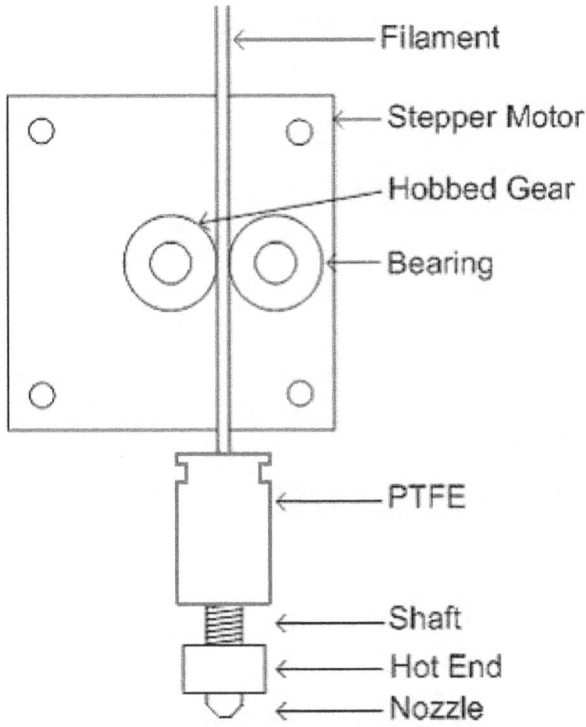

Filament

Stepper Motor

Hobbed Gear

Bearing

PTFE

Shaft

Hot End

Nozzle

I tried adding a spring to the first extruder. It helped a little bit by reducing the amount of unwanted filament strings that jumped across the open gaps. The spring did not help the strength of the parts very much as the hot end was still not able to extrude fast enough. This resulted in a thin or even missing spots in the printed part.

Then I purchased an "extruder upgrade kit" expecting it to fit my existing extruder. The kit required a new hobbed drive gear. Worse yet it actually did not fit on my existing extruder. The metal bar at the bottom of my old extruder was too thick and the hole in the metal bar for the hot end did not line up with the new one. Here is what the extruder upgrade ad looked like on eBay.

Here is a picture of the MK7 Extruder drive gear. I actually needed two of these, one for each of the new extruders.

New MK7 Stainless Steel Extruder Drive Gear Hobbed Gear For Reprap 3D Printer
(261567139009)

ITEM PRICE:
US $1.72

Here is a picture of an "All metal" Hot end. It needs a cooling fan but it runs very cool. In fact when a fan is attached you can generally touch any of the cooling fins while it is running and only the bottom two or three are hot to the touch.

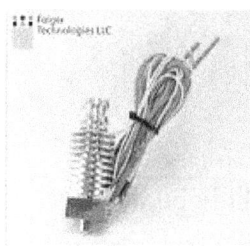

All Metal hotend for ABS PLA For E3D or j-head type filament extruder 1.75mm
(231274375828)

ITEM PRICE:
US $30.14

Here is a picture of the "J-Head" Hot end for the 3D printed extruder. The "J" head is cooler at the top because it is built out of an insulator. That way the heat will not melt the cold end where it is mounted. This is actually the wrong part because it had an adapter at the top that was to be used with a long tube that then allows the cold end to be mounted somewhere besides directly above the hot end.

New J-Head MKIV MKV Hot End 0.4mm Nozzle 1.75mm Filamnet for 3D Printer RepRap
(111262000442)

ITEM PRICE:
US $23.99

Next I built a mounting bracket for the second extruder. It was made out of a piece of 3/4 inch by 1/2 inch "L" shaped aluminum about 2.5 inches long. It could have been a little longer. I then drilled two holes in the aluminum for mounting the stepper motor, two mounting holes to connect to the X platform, and a larger hole for the hot end. I drilled and filed that hole out to about 1/2 inch diameter then cut the metal to bring that hole to the outside edge so the groove in the hot end could slide into that opening..

There were two additional small "L" shaped metal brackets, one is used to hold the 15 pin plug board and the second held the cooling fan in place. I also needed to make a six pin to four pin adapter for the stepper motor. This extruder really worked great. It worked so well that I hated to remove it to test out the third extruder.

Here is a picture of the second extruder with the home made aluminum mounting bracket. This extruder works with the PTFE hot end as well.

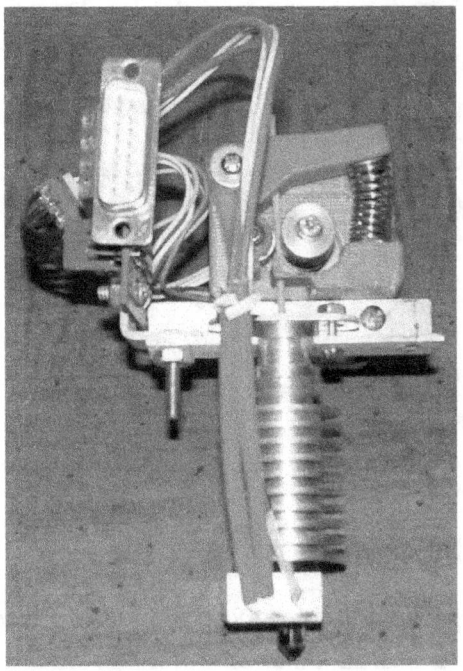

Another thing I tried to do to fix the extruder problem was print out my own extruder. However I had three problems in getting this printed extruder to work. First I did not have the hobbed drive gear at that time. I also did not know the correct size of the bearing that would properly fit into the extruder. When I finally did get some assorted bearings there was one bearing that was too big and one bearing that was too small. This extruder design required a "J head" type of hot end, something I did not have at that time. So I ordered a J Head hot end and waited.

This next picture is of the third extruder. I finally got it to work with a 1/4 inch inside by 3/4 inch outside diameter bearing. I had to file out the opening to make it slightly larger so that the bearing would fit. A 1/4 inch bolt was then cut off to just short of 3/4 inches long to go through the center of the bearing. I used two 4-40 by one inch screws with nuts to hold the hot end in place. The tension is controlled by a 1.75 inch long 6-32 screw and nut. The nut had to be melted into place as the opening was not quite big enough for it.

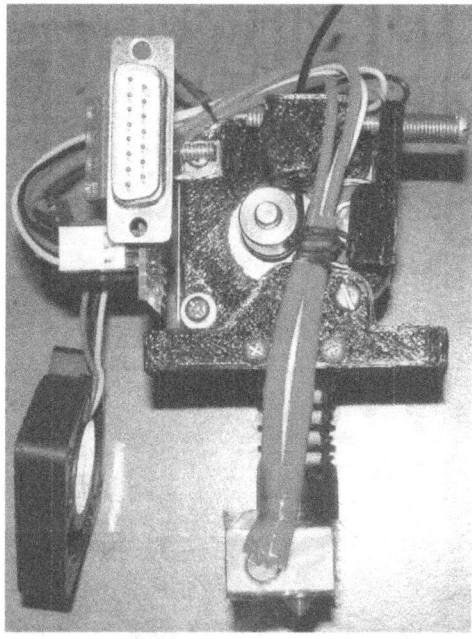

This is a chart comparing the three extruders that I tested out:

Test	Extruder One Heat Sink Above	Extruder Two All Metal Hot End	Extruder Three MK4 J-Head
Overall Height	Shortest	Tallest	Medium Height
Does it leak?	Leaks	Leaks	No Leaks

Does it jam?	Yes	No Jams	No Jams
Changeable Nozzle Size?	Yes	Yes	No Part of Hot End
Did it work?	Poor Results	Great Results	Acceptable Results*

* Extruder three might have performed better with a stronger spring.
Here is a picture showing the three "cold" ends. The left is from extruder one, the second is from extruder two and the third is from extruder three.

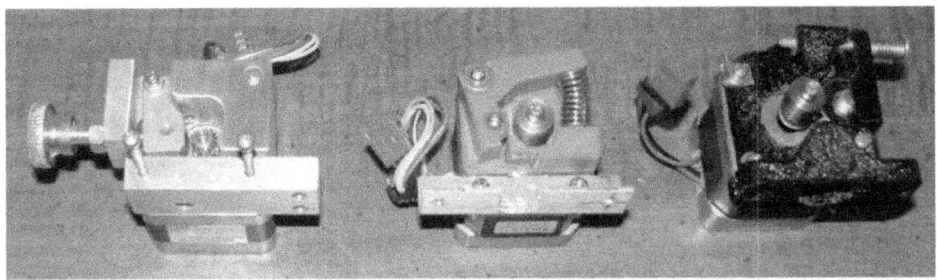

Here is a picture of the three "hot" ends, notice the different lengths. They are arranged by height. The left one is from extruder one, the middle one is from extruder three, and the right one is from extruder two. The hot ends and the cold ends of extruders two and three can be interchanged and were swapped at one point to help verify the test results.

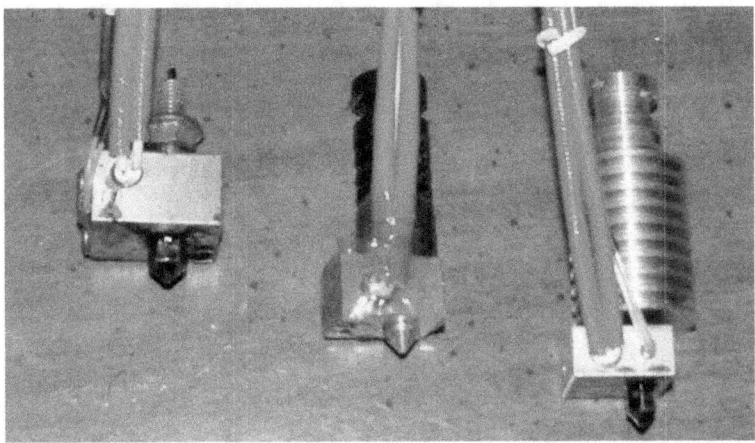

Here is a picture of three 20mm by 20mm test blocks that I printed out to test the extruders. The left test block was printed with extruder number one. The middle was printed with extruder two. The right one was printed with extruder three. The first is a mess (you can actually see light through it) the second is "perfect", and the third is "acceptable".

My conclusion is that extruder two was by far the best. Its drawbacks of loosing height in the Z axis and leaking plastic from its fittings were not as bad as its ability to extrude faster than anything else that I have tested. I could run Marlin at full speed with no problems.

Extruder three was very usable. It does not leak and did not sacrifice as much of a loss in the Z axis. However it sometimes did not keep up at full speed. That might have been because the spring was not strong enough in the cold end. Slowing the printer down slightly resolved the problems.

My tested were run on a home made 3d printer modeled after the Mendel90. I did not verify the temperature of the hot ends with an external temperature meter. I used the same settings in Pronterface and Slic3r with all three of the extruders. Tests were run using PLA filament at 185 degrees centigrade.

Glossary

ABS - Acrylonitrile Butadiene Styrene - plastic filament used in the 3D printing process.

Arduino - Open source single board computer. Uses Atmel processor with built in Analog and digital I/O.

Build Plate/Platform - Moving surface on which the 3D printer prints.

Dual Extruder - 3D printer with two extruders. Gives the ability to print something in two colors at once.

Extruder - Device that melts the filament and deposits the melted material.

Filament - plastic material that is melted and placed by a 3D printer to make objects.

FFF - Fused Filament Fabrication - process of manufacturing in which a filament is melted and deposited into layers to make a 3D object.

Infill - Interior of a 3D object. Usually it is a mesh designed to be solid yet conserve on plastic.

Layer Height - Thickness of a layer of an object being printed usually a little smaller than the nozzle size.

Nozzle - The tip of the hot end of the extruder. Usually a fraction of a millimeter such as .4mm. Controls the size of the output of the extruder.

Hot End - The tip of the extruder that is heated to melt the filament.

Heat Bed - The surface that the 3d object is printed on to. It is heated so the object will not cool too fast and warp.

Linear Bearings - Ball bearings designed to slide along smooth rods.

Mega - Biggest one of Arduino single board computers.

Overhang - Part of a 3D printed object that had nothing below it to support it. This can be done but it is tricky to do.

PLA - Polyactic Acid - Filament that is plant based. It is biodegradable.

RAMPS - Reprap Arduino Mega Pololu Shield - Circuit board with stepper controllers and other I/O for a CNC or 4D printer built in.

Shield - Circuit board that plugs into Arduino Signle Board computers.

Slice - Cut the object to be created into layers to be made.

Stepper Motor - Motor designed to move in small steps under a controller.

StepStick - Single board stepper motor driver.

Thermistor - Temperature sensing resistor.

Bibliography

Here are some of the books that I have thoroughly studied to help me to create these 3D printers.

Mendel 90 Dibond Kit instructions.
No copyright or author is given.

Prusa Mendel Visual Instructions (Standard Resolution)
Original Authors:
    Prusajr (design), Kliment (maintenance and documentation)
Author of this Document:
    Gary Hodgson (http://garyhodgson.com/reprap)
http://reprap.org/wiki/Prusa as of 20th March 2011

Reprap_Prusa_Mendel_Build_Manual_MASTER_V2_Reduced
    Copyright 2012 NextDayReprap All registered trademarks belong
    to their respective owners, any unauthorized copying and
    redistribution is strictly prohibited. E&OE

Building Your Own 3D Printer, An introduction
    By: Steven Devijver
    Copyright © 2011 Steven Devijver

Aluminum I3 packing list
Copyright of this manual belongs to the Shenzhen GETECH CO., LTD.

Geeetech Prusa i3 Metal Building Instruction
Copyright of this manual belongs to the Shenzhen GETECH CO., LTD.

Prusa i3 build manual v1.0
The Prusa i3 (iteration 3) is the newest and current 3D Printer design by RepRap Core Developer Prusajr.

Marlin User Guide
Author: Bart Meijer

I could never list all of the web sites that I have used in creating this book. Here is just a few of the ones that I have added to my favorites.

This site has lots of good pictures and close ups of building a Mendel90.

http://bryanseger.wordpress.com/category/mendel90

This site has a lot of Mendel90 stuff too; he is "Nop Head" the designer.

http://hydraraptor.blogspot.co.uk/

Here is the home of the Mendel90 3d printer.

http://mendel90.blogspot.com/

How about a aluminum Mendel90? Here is one.

http://frankieflood.blogspot.com/2012/07/aluminum-mendel90.html

Here is a laser cut Mendel90, it uses less of the printed parts.

http://blog.think3dprint3d.com/2013/10/mendel90-lasercut-overview.html

Here are the files for printing the parts for making the Mendel90.

http://www.thingiverse.com/thing:17826/#files

www.ingramcontent.com/pod-product-compliance
Lightning Source LLC
Chambersburg PA
CBHW080304180526
45167CB00006B/2659